T0135578

Moustapha M. Diallo

Down regulation of genes involved in
sphingolipid metabolism in cultured mammalian
and Drosophila S2 cells by RNA interference

λογος

Bibliografische Information der Deutschen Nationalbibliothek

Die Deutsche Nationalbibliothek verzeichnet diese Publikation in der
Deutschen Nationalbibliografie; detaillierte bibliografische Daten sind
im Internet über http://dnb.d-nb.de abrufbar.

©Copyright Logos Verlag Berlin GmbH 2011
Alle Rechte vorbehalten.

ISBN 978-3-8325-2823-2

Logos Verlag Berlin GmbH
Comeniushof, Gubener Str. 47,
10243 Berlin
Tel.: +49 (0)30 42 85 10 90
Fax: +49 (0)30 42 85 10 92
INTERNET: http://www.logos-verlag.de

1.   Abbreviations

Ac	Acetate
Ago	Argonaute
Amp	Ampicillin
Arg	Arginin
AP	Alkaline phosphatase
APS	Ammonium peroxodisulfat
ATP	Adenosine triphosphate
B	Biotin
b	Bases, nucleotides
• -hex A	• -Hexosaminidase A
• -Hex A	• -Hexosaminidase A gene
Bp	Base pairs
BSA	Bovine serum albumin
C	Cystein
cDNA	Copy- desoxyribonucleic acid
Ci	Curie
cpm	counts per minute
Cys	Cysteine
dATP	Desoxyadenosine triphosphate
dCTP	Desoxycytidine triphosphate
ddATP	Didesoxyadenosine triphosphate
ddCTP	Didesoxycytidine triphosphate
ddGTP	Didesoxyguanosine triphosphate
ddTTP	Didesoxythymidine triphosphate
DEPC	Diethylpyrocarbonate
DIG	Digoxigenin
dGTP	Desoxyguanosintriphosphat
DMEM	Dulbecco´s Modified Eagle´s Medium
DMF	Dimethylformamide
DMSO	Dimethylsulfoxide
DNA	Desoxyribonucleic acid
DNAse	DNA cleavage enzyme
dNTP	Desoxyribonucleotide tri-phosphate
ds	double stranded
dsRNA	double stranded RNA
DTT	Dithiothreitol
dTTP	Desoxythymidine triphosphate
EDTA	Ethylendiamine tetraacetat

ELISA	Enzyme-linked immunosorbent assay
ER	Endoplasmatic reticulum
EtOH	Ethanol
FSC	Foetal caf serum
FITC	Fluoresceine isothiocyanat
GalCer	Galactosylceramide,
GBA	ß-Glucocerebrosidase
GlcCer	Glucosylceramide,
GM1	Ganglioside GM1
GM2	Ganglioside GM2,
GM3	Ganglioside GM3,
GM2A	GM2 aktivator gene
GM2AP	GM2 aktivator protein
GSP	Gene specific primer
GTP	Guanosine triphosphate
H	Histidine
HEPES	2-[4-(2-Hydroxyethyl)- 1piperazino]-ethansulfon acid
His	Histidine
i	inverse
IPTG	Isopropyl-1-thio-ß-D- galactosid
kb	Kilo Base Pairs
kDa	Kilo Dalton
LB	Luria Bertani medium
lhRNA	long hairpin RNA
Lys	Lysine
mA	Milliampere
MEB4	Mouse melanocyte
MEM	Minimum Essential Eagle Medium
Met	Methionin
NCBI	National Center of bioinformatics
ng	Nanogram
mRNA	Messenger-RNA
NTP	Nucleoside triphosphate
OAS	oligoadenylate synthase
OD	Optical density
ORF	open reading frame
PAA	Polyacrylamide
PAG	Polyacrylamid gel
PAGE	Polyacrylamide gel electrophoresis
PBS	Phosphat buffered saline
PCR	Polymerase chain reaction

PDMP	1-phenyl-2-decanoylamino-3- -morpholinopropanol	shRNA	short haipin RNA
		ss	single stranded
PFA	Paraformaldehyde	SSC	Standard saline citrate
PIPES	1,4-Piperazin -bis (ethansulphon acid)	TAE	Tris-acetate-EDTA buffer
		TE	Tris-EDTA Buffer
pSAP	Prosaposine	TEMED	N, N, N´, N´- Tetramethylen diamine
RISC	RNA Induced Silencing complexe		
RNA	Ribonucleic acid	Tm	Melting temperature
RNAi	RNA interference	Tris	Trishydroxymethylaminometh- ane
RNAse	RNA digesting enzyme		
RNAsin	RNAse-inhibitor	TTP	Thymidine triphosphate
rpm	rounds per minute	ug	Microgram
RT	Room temperature	u	Units
RT-PCR	Reverse-transkription-PCR	UTR	Untranslated region
SAP	Sphingolipid activator protein	V	Volt
SDS	Sodium dodecylsulfate	Vol	Volume
siRNA	small interfering RNA	x-Gal	5-Chlor-4-bromo-3-indolyl-ß- D-galactose

2.   Summary

Elucidating gene function not only represents a crucial interest in fundamental research, but also in understanding the genesis of genetic deseases and in designing therapeutic strategies for treatment. Various methods have been designed and developed to analyze gene function in cell, tissue and organism systems. The most prominent of them are the gene disruption or gene silencing by homologue recombination, the use of antisense oligonucleotides to prevent the translation of messenger RNA into the corresponding protein, the overexpression of a modified, non fuctional version of the protein of interest, the use of antibody directed against the protein of interest. However, these methods are often very laborious and time consumming.

Since one decade, RNA interference (RNAi), which is a phenomenon of sequence specific degradation of mRNA mediated by homologous short interfering RNA (siRNA) or doubble stranded RNA (dsRNA), has become the method of choice to analyse gene function in cells and organisms. According to the cell type or organism in which the analysis is carried out, different RNAi based techniques have been developed and applied. siRNA, shRNA and lhRNA, either generated *in-vitro* or endogenously synthetized by a plasmid are currently used in invertebrates like *C. elegans* and *D. melanogaster*. However, siRNA and shRNA based silencing is predominantly used in mammalian cell systems as compared to long dsRNA or long hairping RNA, which is due to the ability of the latter to trigger the interferon cascade in mammalian cells, leading to apoptotic death.

However, in this work, we could show that RNA polymerase II driven lhRNA expression can also be used in mammalian cells. We design a plasmid based RNAi model to permanently down regulate gene expression in cultured mammalian cells. This system consist in engeneering cells to continously express a lhRNA that targets the homologous mRNA of interest for sequence specific degradation. We speculate that the amount of the synthesized dsRNA will be moderate and regularly processed into siRNAs, so that it cannot accumulate beyond the threshold required to trigger the apoptotic pathway. To get a proof of this principle, we generated mammalian cells stably expressing dsRNAs targeting genes of the sphinglipid metabolism for silencing, namely the glucosylceramide synthase (GCS) and the prosaposine (Psap) genes, as we could compare phenotypes with patient material.

On the other hand, we analysed the function of the schlank gene in Drosophila S2 cells using *in-vitro* synthetized dsRNA. The schlank protein is believed to be involved in ceramide synthesis and in the regulation of lipid homeostasis in *Drosophila melanogaster*.

3. Introduction

3.1 Background

When in the early 40, Oswald Avery and coworkers identified the Deoxyribonucleic acid (DNA) as the biochemical molecule responsible for rendering the R (Rough), non pathogenic bacterial stock, *Diplococcus pneumonae,* into the S (Smooth) pathogenic stock observed by A. Grifith in 1928 when he mixed the first one with the heat killed other stock, they switched at this moment the convulsions that lead to the birth of modern genetic. Soon after, DNA not only reveals to govern the fundamental characteristics of every living entity, but also to be the porter and the bridge of heredity between generations. Every organism is equipped with a species-specific genetic program, with a defined set of genes – the genome. The essential biological and biochemical behaviors of every organism are the result of the execution of this genetic program through its controlled expression. This knowledge profoundly modified the searcher's perception of life, and the genome and its encoded products – RNAs and proteins -, conceived as the motor of life, became the gravitational centre of biological researches. Thus different methods have been developed by the researchers to allow them to subject DNA to *in-vitro* and *in-vivo* experimental manipulations and to influence at will the expression of genetic information. This opportunity gave rise to the most gigantic and most ambitious research project in genetic, aiming at the identification of whole genes of living organisms by sequencing their genome, and establishing their respective roles in the integrated biochemical living system by way of reverse genetic analyses. Whereas forward genetic begins with the mutant phenotype to show the existence of the relevant gene by classical genetics, and finally cloning and sequencing the gene to determine its DNA and protein sequence, reverse genetics, a new approach made possible by recombinant DNA technology, works in the opposite direction. It starts from a protein or gene for which there is no genetic information and then works backwards to induce a silencing of gene expression, finishing with the resulting phenotype. Reverse genetic studies, commonly called functional genomics is the way to meet the challenge imposed by a steadily, from sequencing projects of whole genomes, accumulating genetic data to decipher the genetic information they encode for. Meanwhile the genetic landscape of several organisms has been deciphered, including *C. elegans, D. melanogaster* and human; and sequencing projects of others are going on.

At purpose of functional genomics, several methods have been developed by the reresearchers, and used to stop the flow of genetic information at either the DNA or the RNA level, or also at the protein level. Gene silencing at the DNA level mainly proceeds by isolating the WT gene and subjecting it to *in-vitro* mutagenesis by way of nucleotide substitution, deletion, or insertion. The mutated gene is then put back into cultured cells or into the organism of interest where it replaces the functioning gene through homologue recombination (Arenz C., and Schepers U., 2003). This method has been currently used to silence gene in cultured cells and to generate knock out animals for specific genes. Particularly important is also the adapted method of homologue recombination enabling time and tissue specific conditional knock out that is actually used to analyze the function of embryonic lethal genes in adult organisms.

The other way currently used at purpose of functional genomics studies is the anti sense technique based on the use of DNA or RNA oligonucleotide complementary to defined sequences of the messenger encoded by the gene of interest (Arenz C., and Schepers U., 2003). The complementary base pairing between DNA or RNA oligonucleotide and the messenger negatively impacts on the translation of the later into the corresponding protein and causes the mutant phenotype to appear (Arenz C., and Schepers U., 2003). As well other techniques use specific molecules – like Antibodies, inhibitors, or haptamers to achieve gene-silencing *in-vitro* at the protein level. These molecules specifically interact with the target protein preventing it to play its role by building an inactive complex. These different methods have been more or less successfully used to analyze the role of a wide range of genes in cultured cells and in different model organisms including mouse.

3.2 RNAi

RNA interference (RNAi), defined as a process of dsRNA induced gene silencing based on the sequence specific degradation of homologous mRNA (Fire et al., 1998; Schepers and Kolter 2001) is the most exciting discovery in biology within the last two decades. 1995 Guo and Kemphues, in an anti sense experiment they were performing to down regulate the par1 polarity gene in the worm *C. elegans*, surprisingly observed that both the experimental probe containing the anti sense and the control probe containing the sense RNA strand caused the same silencing effect. 1998, Timons and Fire carried out the first intended RNAi experiment in *C. elegans* by introducing simultaneously either the sense and anti sense RNA strands, and obtained a high interference with the expression of their target gene. They proposed such a silencing effect – to which they gave the name RNAi – to be mediated by both RNA strands as double stranded RNA (dsRNA), and underlying another more complicated molecular pathway than the classical anti sense silencing (Timmons and Fire 1998).With this experiment, they brought the explanation to the observation made by Guo and Kemphues, whose probes must have been cross contaminated.

Intensive investigations of the RNAi process have allow the researchers to show it beeing related to well known phenomenons of post transcriptional gene silencing occuring in plants (PTGS), Drosophila (*co-suppression*) and in *N. crassa (Quelling)* that originally evolved as a genome defence mechanism against invading genetic elements like RNA viruses and transposons (Zamore, Tuschl et al. 2000) Nowadays, much evidence indicates that RNAi is an evolutionarily, beyond all eukaryotes, conserved molecular pathway involved in the regulation of physiological processes ranging from chromatin dynamics to gene regulation during development.

3.2.1 Mecanism of working

Soon after the discovery of RNAi, researchers challenged the question relative to its mechanism of working. And this question has primarily been addressed using genetics in *C. elegans*, plants and fungi, and by biochemical studies using *Drosophila* cell extracts.

Figure 3-1 outlines the current model of the RNAi mechanism. The proper triggers of the RNAi pathway are oligonucleotide fragments of 21-23 bps, commonly called short interfering RNA (siRNA), which can be generated by Dicer from dsRNA introduced into the cells or by chemical (Zang et al., 2004, Zamore et al., 2000; Elbashir et al., 2001) or biochemical synthesis. Dicer is complexed with the TAR-RNA-binding proteins (TRBP), PACT, and Ago2, which are RNA-binding proteins. SiRNAs have characteristic 2 nt 3' overhangs allowing them to be recognized by the machinery of RNAi, therefore leading to the degradation of the target mRNA. The 2 nt 3' overhangs of the siRNA molecules are critical for them to be handed-off to the next multifunctional protein complex – the RNA-induced silencing complex (RISC) by the RNA-binding proteins TRBP, PACT, and Ago2 (Lee et al., 2006). The core components of RISC are the Ago family members showing a domain for cleavage activity (Mater et al., 2004; Liu et al., 2004). While the siRNAs loaded into RISC are double stranded, Ago2 cleaves and release the passenger strand, which leads to the activated form of RISC containing a single stranded guide RNA molecule that directs the specificity of the target recognition by intramolecular base pairing (Tang et al., 2005), and therefore to its degradation (Figure 3-1, left panel). Differential thermodynamic stabilities of the ends of the siRNAs seem to govern the selectivity of strand loading into the RISC (Schwarz et al., 2003; Khvorova et al., 2003).

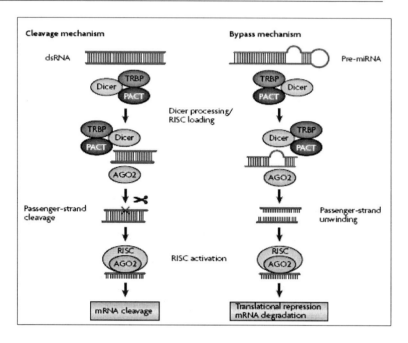

Figure 3-1 Mecanisms of RNAi silencing: depending on whether the two complementary strands of the dsRNA perfectly or partially match the one to the other, silencing can lead to mRNA degradation (left panel) or to translational repression (right panel) respectively.

The native pathway of RNAi with respect to the regulation of gene expression is mediated by microRNAs (miRNAs). These are structurally similar to the siRNAs and as well develop their silencing effect by building a complexe with RISC and binding to the 3' untranslated regions (UTRs) of target sequences via short stretches of homologie termed seed sequences (Bartel et al., 2004). Depending on the degree of complementarity between miRNA and the target mRNA, the primary mechanism of action of miRNAs can result in translational repression (mismatched complementarity) or in message degradation (complete complementarity) (Bagga et al., 2005). MiRNAs douplexes possess incomplet Warson-Crick base pairing, and the anti sense strand cannot be chosen by cleavage of the passenger strand as it is for siRNAs; instead, it must be chosen by an alternativ mechanism (Leuchener at al., 2006). miRNAs are expressed as long primary transcripts (pri-miRNAs), which are processed within the nucleus into 60-70 bp hairpins by the micro processor complex consisting of Drosha and DGCR8 into pre-miRNAs (Lee et al., 2003; Han et al., 2004). These latter are transported into the cytosol, where they are further processed by Dicer into miRNAs, which are unwounded, whereby one of the strands is loaded into RISC, presumably via interaction with one of the Dicer accessor proteins (Lee et al., 2006). The activated RISC recrutes the target mRNA by intramolecular base pairing with the miRNA strand leading to translational repression in case of complete complementarity or to target degradation in cas of mismatched complementarity (Figure 3-1, right panel).

3.1.2 RNAi as a tool in functional genomic analysis

Soon, reresearchers exploited the practical opportunity the RNAi pathway offers in experimental biology to study gene function. Due to the facility for it to be carried out and the short time within which results can be available, RNAi based gene silencing has become the standard procedure in functional genomic analysis, therefore replacing the above cited techniques, which are mostly afflicted with time-consoming crosses and trickly cloning. RNAi based analyse of gene function was primarily applied on model organisms like *Drosphila melanogaster, C. elagans* and plants (Daneholt, B., 2007; Kamath R. and Ahringer J., 2003; Boutros et al., 2004). In worms and fly embryos, RNAi is delivered by the injection of the dsRNA (Fire et al., 1998). As well, systemic RNAi can be induced in worms by simply feeding them bacteria that express an appropriated dsRNA (Timons at al., 2001). Into cultured *Drosophila* cells, dsRNA can be delivered by simple transfection and also in other cells by simply adding it to growth media in the absence of the transfection reagenz (Caplen at al., 2000; Clemens et al. 2000). Further, dsRNA can be delivered to invertebrates by generating transgenic organisms expressing hairpin RNA *in vivo* from general or tissue specific promoters (Kennerdell et al., 2000; Timmons et al., 2003). In plants, dsRNA or hairpin transgenes can be delivered using standard transformation procedures, but also by incorporating target sequences into a virus that encodes an RNA- dependent RNA polymerase (Waterhouse and Helliwell 2003).

In mammalian cells, the introduction of long dsRNA into the cytoplasm activates the interferon response; a cellular anti-viral response that leads to non-specific translational inhibition and, potentially, apoptotic cell death (Williams et al., 1998; Reynolds et al., 2006). Exceptionally, mouse oocytes and cells from early mouse embryos lack this reaction to exogenous dsRNA and have constituted therefore a first common model system for RNAi based gene-knockdown effects in mammals (Stein et al., 2005).

To avoid the induction of the interferon pathway by long dsRNA, the RNAi based gene knock down in cultured mammalian cells uses generally siRNAs of 21 nucleotides (Elbashir et al., 2001). siRNAs can be chemically synthesised, *in-vitro* transcribed by T7 polymerase (Yu et al., 2002; Donze et al., 2002; Paddison et al., 2002) or produced by enzymatic cleavage from *in-vitro* synthesised dsRNAs (Myers et al., 2003) and then transfected into tissue culture cells using standard procedures. However, chemical synthesis of siRNAs is a relatively expensive process and in vitro processing is time consuming, the products will likely need to be size purified and both can only produce transient silencing. To overcome these limitations, many laboratories have developed vectors or lentiviruses that express short hairpin RNAs (shRNAs) or long hairpin RNAs (lhRNAs) for the purpose of long lasting silencing effect. These plasmid or lentiviral based systems produce RNAs with 19 to 29 or several 100 base pair stems that are processed by Dicer in vivo to produce mature siRNAs (Paddison et al., 2002). In this work, we developed a plasmid based RNAi system to down regulate the glucosylceramide synthase (GCS) and the prosaposine (Psap) genes in cultured mammalian cell lines. These plasmids were designed to express an inverted RNA repeat of 1600 bp that fold back to produce long hairpin RNAs of 800 bps targeting the respective mRNA of glucoceramide synthase (GCS) and prosaposine (PSAP) for sequence specific degradation (Diallo et al., 2003). On the other hand, we analysed the the role the *schlank* gene in cultured Drosphila S2 cells by using in-vitro generated dsRNA of 800 base pairs.

3.3 Sphingolipids

3.3.1 Background

Sphingolipids are a class of lipids derived from the aliphatic amino alcohol sphingosin. These compounds play important roles in signal transduction and cell recognition. The long-chain bases, sometimes simply known as sphingoid bases, are the first non-transient products of *de novo* sphingolipid synthesis in both yeast and mammals. These compounds, specifically known as phytosphingosine and dihydrosphingosine, are mainly C_{18} compounds, with somewhat lower levels of C_{20} bases. Ceramides and glycosphingolipids are *N*-acyl derivatives of these compounds. The sphingosine backbone is either O-linked to a (usually) charged head group such as ethanolamine, serine or choline, or is amide-linked to an acyl group, such as a fatty acid. Ceramide is the fundamental structural unit common to all sphingolipids.

There are three main types of sphingolipids, differing in their head groups:

The sphingomyelins that have a phosphorylcholine or phosphoroethanolamine molecule with an ether linkage to the 1-hydroxy group of a ceramide

The glycosphingolipids, which are ceramides with one or more sugar residues, joined in a β-glycosidic linkage at the 1-hydroxyl position. According to the nature of the substituents on their head group they are classified in cerebrosides with a single glucose or galactose at the 1-hydroxy position, and sulfatides, which are sulfated cerebrosides.

The Gangliosides that have at least three sugars, whereby one of which must be sialic acid.

3.3.2 Functions in mammals

Sphingolipids are commonly believed to protect the cell surface against harmful environmental factors by forming a mechanically stable and chemically resistant outer leaflet of the plasma membrane lipid bilayer. Certain complex glycosphingolipids were found to be involved in specific functions, such as cell recognition and signaling. The first feature depends mainly on the physical properties of the sphingolipids, whereas signaling involves specific interactions of the glycan structures of glycosphingolipids with similar lipids present on neighboring cells.

Relatively simple sphingolipid metabolites, such as ceramide and sphingosine-1-phosphate, have been shown to be important mediators in the signaling cascades involved in apoptosis, proliferation, and stress responses (Spiegel and Milstien 2002; Hannun et al. 2002). Ceramide-based lipids self-aggregate in cell membranes and form separate phases less fluid than the bulk phospholipids. These sphingolipid-based microdomains, or "lipid rafts" were originally proposed to sort membrane proteins along the cellular pathways of membrane transport. At present, most of the research focuses on the organizing function during signal transduction (Brown and London 2000).

3.3.3 Sphingolipids biosynthesis

Sphingolipids are synthesized in a pathway that begins in the ER and is completed in the Golgi apparatus (Figure 3-2); later, these lipids are enriched in plasma membrane and in endosomes, where they perform many of their functions (Futerman et al. 2006). Transport occurs via vesicles and monomeric transport in the cytosol. Sphingolipids are virtually absent from mitochondria and the ER, but constitute a 20-35 molar fraction of plasma membrane lipids (van Meer et al. 2002). The de *novo* sphingolipid synthesis begins at the endoplasmic reticulum with the formation of 3-keto-dihydrosphinganine by serine palmitoyltransferase (Merrill 1983). The preferred substrates for this reaction are palmitoyl-CoA and serine. However, studies have demonstrated that serine palmitoyltransferase has some activity towards other species of fatty acyl-CoA (Merrill et al. 1984) and alternative amino acids (Zitomer et al. 2009), and the diversity of sphingoid bases has recently been reviewed (Pruett et al. 2008). In the next step, 3-keto-dihydrosphinganine is reduced to form dihydrosphinganine (Figure 3-2). Dihydrosphingosine is acylated by a (dihydro)-ceramide synthase, such as Lass1p or Lass2p (also termed as CerS), to form dihydroceramide (Pewzer-Jung et al. 2006), which is desaturated to form ceramide (Causeret et al. 2005) (Figure 3-2). Ceramide is subsequently transported to the Golgi by vesicular trafficking or the ceramide transfer protein CERT, where it can be further glycosylated by glucosylceramide synthase to form glucosylceramide, or galactosylceramide synthase to form galactosylceramide, the two precursors of complex glycosphingolipids. Alternatively, it may be phosphorylated by ceramide kinase to form ceramide-1-phosphate, or it can be converted to sphingomyelin by the addition of a phosphorylcholine headgroup by sphingomyelin synthase. Diacylglycerol is generated by this process.

3.3.4 Glycosphingolipid biosynthesis

The synthesis of Glucosylceramide is important as the first synthetic step of numerous GSLs. GCS (UDP-glucose: ceramide glucsyltransferase, UGCG) transfers a glucose residue from UDP-glucose to ceramide and produces glucosylceramide (Figure 3-2) (Schulte and Stoffel, 1993; Stahl et al., 1994). This step tightly regulates the production of all upstream GSLs (Coetzee et al., 1996).

Sequence analysis suggests that GCS has a type III membrane protein structure with N-terminal signal anchor sequence and a long cytoplasmic tail (Futerman and Pagano, 1991; Nozue et al., 1988; Ichikawa et al., 1994 and 1996; Coste 1986, and Jeckel et al. 1992). The structure of the enzyme ist quite unique since all other glycosytransferases involved in GSL synthcsis arc localised to the luminal side of the Golgi apparatus or the endoplasmatic reticulum (ER). The orientation of the enzyme indicates that its GlcCer product has to be translocated to the luminal side of the Golgi membrane for further glycosylation. This translocation is presumably mediated by a protein, a flippase, because lipids with polar groups rarely transfere between bilayers spontaneously (Lannert et al. 1994; Burger et al., 1996). However, there is still no experimental evedence for a flippase for GlcCer. As indicated above, sphingolipids, particularly ceramide, have been shown to play an important role in signal transduction, regulation and cell homeostasis, and sphingolipid metabolism is highly regulated in response to extracellular and intracellular signals and stimuli, (Hannun 1994; Kolesnick 1994). The reaction catalysed by the GCS might be essential regulatory factor controlling the cellular level of this bioactive lipid molecule. In fact, the activity of the GCS is stimulated by a number of means that increase ceramide concentrations, such as the addition of short-chain ceramides and the treatment with bacterial sphingomyelinase, endoglycoceramidase, or inhibitors of GCS (Abe et al., 1996; Ito and Komori 1997). As a potent bioactive messenger molecule, the elevation of ceramide concentration needs to be transient or it might cause adverse cellular effects. Although ceramide can flip-flop freely to either side of the membrane bilayer, the cytosolic location of GCS is advantageous for the control of cytosolic ceramide concentration. Other phenomena are as well suggestive of important roles for ceramide glucosylation. It was demonstrated that GlcCer accumulates in multidrug-resistant cancer cells and that ceramide glucosylation is associated with the multidrug-resistant state (Lavie et al. 1996 and 1997). Further the enzymatic activity of GCS is regulated developmentaly in the course of epidermal development and differentiation (Sando et al., 1996).

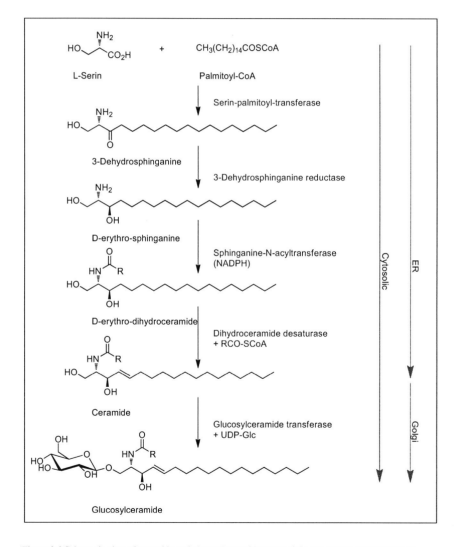

Figure 3-2 Schematic view of ceramide and glucosylceramide synthesis in the cellular compartments.

3.3.5 Sphingolipids degradation

The lysosomal degradation of glycosphingolipids is a sequential pathway that starts with the stepwise release of monosaccharide units from the nonreducing end of the oligosaccharide chain (Figure 3-3). These reactions are catalyzed by exohydrolases with acidic pH-optima. Several of these enzymes need the assistance of small glycoprotein cofactors called Sphingolipid Activator Proteins or SAPs (Sandhoff et al. 2001).

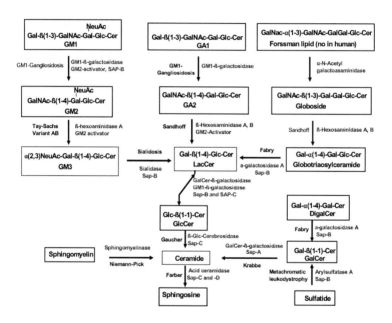

Figure3-3. Degradation of selected sphingolipids in the lysosomes of the cells (modified from Kolter & Sandhoff 1998). The eponyms of individual inherited diseases (shown in red) are given. Activator proteins required for the respective degradation step in vivo are indicated. Variant AB: AB variant of GM2 gangliosidosis (deficiency of GM2-activator protein).

Two genes encoding for SAPs have been identified and charaterised: one encodes the GM2-activator protein, the other encodes the Sap-precursor protein, also called prosaposin (Sandhoff et al. 2001), which is object of this study. In addition to enzymes and activator proteins, low cholesterol content, and the presence of the negatively charged lysosomal lipid bismonoacylglycerophosphate (BMP) are as well important for the degradation of sphingolipids.

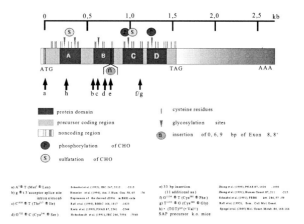

Figure 3-4 Organisation of the full-length cDNA of SAP-precursor (Holtschmidt at al., 1991).
Localization of identified mutations is indicated as follows: a, A1T (MIL) (Schnabel et al. 1992); b, g-t (3'
acceptor splice site intron e/exon 6) (Henseler et al., 1996a); c, 650 > T (T217I) (Kretz et al., 1990; Rafi et al.,
1990); d, 722G > C (C 241S) (Holtschmidt et al., 1991b); e, 33 bp insertion (11 additional aa) (Zang et al., 1990,
1991); f, 1154G > T (C385F) (Schnabel et al., 1991); g, 1155T > C (C385G) (Rafi et al., 1993); h, 643A > C
(N215H) (Wrobe et al., 2000)

The prosaposin (Figure 3-4) is a lysosomal protein of 70Kd that is proteolytically processed to

four homologous mature proteins, Saps A-D, or saposins A-D (Fujibayashi and Wenger

1986a, b; Kishimoto et al. 1992; O'Brien and Kishimoto 1991). These saposins act on the

intra-endosomal/intra-lysosomal membrane pool and lead to the selective degradation of

membrane lipids without impairment of lysosomal integrity. The stepwise cleavage of the

hydrophilic head groups from glycosphingolipids ultimately generates sphingosine, fatty

acids, monosaccharides, sialic acids, and sulfate. These final degradation products are able to

leave the lysosome and are introduced into the salvage pathway. Members of other

glycosphingolipid series enter the degradation pathway of gangliosides at the

lactosylceramide stage. Glycosphingolipids of the gala-series, but also sphingomyelin, are

degraded to ceramide.

Non-glycosylated sphingolipids, such as ceramide and sphingomyelin (Goni & Alonso 2002),

have non-lysosomal degradation steps that apparently do not need the assistance of an

activator protein. A cytoplasmic glucosylceramide-cleaving enzyme, contributes to the

degradation of the cytoplasmic glucosylceramide pool (van Weely et al. 1993).

Inherited deficiency of either lysosomal enzymes or SAPs leads to the accumulation of

nondegradable membranes within the lysosomal compartment and to the development of sphingolipid-storage diseases (Kolter & Sandhoff 1998, Suzuki & Vanier 1999, Platt & Walkley 2004, Winchester 2004) (Figure 3-3)

The prosaposin (Figure 3-4) is a glycoprotein detected mainly in uncleaved state in brain, heart and muscle, whereas mature Saps are mainly found in liver, lung, kidney and spleen. The Sap-precursor also occurs in body fluids such as milk, semen, cerebrospinal fluid, bile, and pancreatic juice. It is either intracellularly targeted to the lysosomes via mannose-6-phosphate receptors or by sortilin (Lefrancois et al. 2003), or it can be secreted and endocytosed by mannose-6-phosphate receptors, low-density lipoprotein receptor-related protein (LRP) (Hirschberger et al. 1998.) To date, two different mutations in four human patients have been reported that lead to a complete deficiency of the whole prosaposine, and consequently of all four Saps. Because the prosaposine is efficiently proteolytically processed within the acidic compartments, it can be assumed that the unprocessed protein plays no role in membrane digestion. In human patients with prosaposine deficiency, and also in the prosaposine knockout mice (Fujita et al. 1996), a simultaneous storage of many sphingolipids, including ceramide, glucosylceramide, lactosylceramide, ganglioside GM3, galactosylceramide, sulfatides, digalactosylceramide and globotriaosylceramide, accompagnied by a dramatic accumulation of intra-lysosomal membranes is observed (Figure 3-3). This storage can be completely reversed by exogenous treatment with human prosaposine, as demonstrated in prosaposin-deficient fibroblasts (Burkhardt et al. 1997).

Although the four Saps share a high degree of homology and similar properties, they act differently and show different specificities.

Sap-A. Sap-A is involved in the degradation of galactosylceramide by the galactosylceramide-ß-galactosidase in vivo. This is demonstrated by the phenotype of mice carrying a mutation in the Sap-A domain of the Sap-precursor. These mice accumulated galactosylceramide and suffer of a late onset form of Krabbe disease (Matsuda et al. 2001). To date, only one human patient has been reported with an isolated defect of Sap-A (Spiegel et al. 2005.)

Sap-B. Sap-B was the first activator protein identified (Mehl & Jatzkewitz 1964). It is essential for the degradation of sulfatide by arylsulfatase A and of globotriaosylceramide and digalactosylceramide by α-galactosidase A in vivo. This has been demonstrated in patients with Sap-B deficiency, where these substrates are found in the urine (Li et al. 1985). Sap-B is also essential for the degradation of other glycolipids (Li et al. 1988). For example, it cooperates with the GM2-activator protein in the degradation of ganglioside GM1 (Wilkening et al. 2000). Similar to GM2-activator, Sap-B acts as a physiological detergent but shows a broader specificity. The crystal structure shows a shell-like homodimer that encloses a large hydrophobic cavity (Ahn et al. 2003). The monomers are composed of four amphipathic α-helices arranged in a long hairpin that is bent into a simple V-shape. As in the GM2-activator, there are two different conformations of the Sap-B dimmers, and a similar mechanism for its action has been proposed: The open conformation should interact directly with the membrane, promote a reorganization of the alkyl chains, and extract the substrate accompagnied by a change to the closed conformation. Thus the substrate could be exposed to the enzyme in a water-soluble activator-lipid complex (Fischer & Jatzkewitz 1977), consistent with the previous observation that Sap-B can act as a lipid-transport protein (Vogel et al. 1991). The inherited defect of Sap-B leads to an atypical form of metachromatic leukodystrophy, with late infantile or juvenile onset (Kretz et al. 1990). The disease is characterized by accumulation of sulfatides, digalactosylceramide, and globotriaosylceramide (Sandhoff et al. 2001).

Sap-C. Sap-C is a homodimer and was initially isolated from the spleen of patients with Gaucher disease (Ho & O'Brien 1971). It is required for the lysosomal degradation of glucosylceramide by glucosylceramide-ß-glucosidase (Ho & O'Brien 1971). In addition, Sap-C renders glucosylceramide-ß-glucosidase more protease resistant inside the cells (Sun et al. 2003). The solution structure of Sap-C (de Alba et al. 2003) consists of five tightly packed α-helices that form half of a sphere. All charged amino acids are solvent-exposed, whereas the hydrophobic residues are contained within the protein core. In contrast to the mode of action of the GM2-activator and of Sap-B, Sap-C can directly activate the glucosylceramide-ß-glucosidase in an allosteric manner (Ho & O'Brien 1971; Berent & Radin 1981; Fabbro & Grabowski 1991). Sap-C also supports the interaction of the enzyme with the substrate embedded in vesicles containing anionic phospholipids, and Sap-C is able to destabilize these vesicles (Wilkening et al. 1998). Binding of Sap-C to phospholipids vesicles is a pH-controlled, reversible process (Vaccaro et al. 1995). Sap-C deficiency leads to an abnormal juvenile form of Gaucher disease and an accumulation of glucosylceramide (Christomanou et al. 1986; Schnabel et al. 1991.)

Sap-D. Sap-D stimulates lysosomal ceramide degradation by acid ceramidase in cultured cells (Klein et al. 1994) and *in vitro* (Linke et al. 2001b). Morover, it stimulates acid sphingomyelinase-catalyzed sphingomyelin hydrolysis, but this seems not to be necessary for the in vivo degradation of sphingomyelin (Linke et al. 2001a). The detailed physiological function and mode of action of Sap-D is unclear. It is able to bind to vesicles containing negatively charged lipids and to solubilize them at an appropriated pH (Ciaffoni et al. 2001).

Sap-D deficient mice accumulate ceramide with hydroxylated fatty acid mainly in the brain and in the kidney (Matsuda et al. 2004.)

3.3.6 Role of saposins in lipid antigen presentation

In addition to their function as enzyme cofactors, SAPs play an important role in the presentation of lipid and glycolipid antigens. It is now established that CD1 immunoreceptors present lipid antigens to T cells. However, these lipids must first be removed from the membranes, in which they are embedded to allow loading of CD1 molecules. The human genome encodes four MHC-I-like glycoproteins (CD1a-d) that presents lipids to T cells. A fifth gene encodes CD1e, which is synthetized as an integral membrane protein and from which a soluble lipid-binding domain is released by proteolysis within the lysosomes of mature dendritic cells (Angenieux et al. 2005)

A possible function of this protein might also be lipid transfer, but this has not been proven to date. The three-dimensional structures of protein-lipid-complexes between human CD1b and two lipids, phosphatidylinositol and ganglioside GM2 (Gadola et al. 2002), and between CD1a and sulfatide (Zajonc et al. 2003) have been reported. There is evidence that SAPs acting as lipid transfer proteins participate in loading process within the acidic compartments of the cells. Antigen presentation by CD1b (Winau et al. 2004), as well as human (Kang et al. 2004) and mouse CD1d (Zhou et al. 2004), have been studied: Human CD1b especially requires Sap-C to present different types of glycolipid antigens. In vitro, all saposins can exchange phosphatidylserin bound to murin CD1d against glycosphigolipids, but with different acitivity. The state of our knowledge in this regard is incomplete. Further research is required to reach a better understanding of the mechanisms involved in this process.

3.4 Drosophila schlank

In addition to their role as structural components and source of energy, lipids and sphingolipids are implicated in various physiological processes occurring in eukaryotic cells, including growth, differentiation, proliferation, transport and survival (Spiegel and Milstien, 2000; Hannun et al., 2001; Acharya and Acharya, 2005). Accordingly, lipid and sphingolipid homeostasis must be tightly controlled and adapted to the strict structural, energetic, and physiological requirements of the cells and living organisms.

As mentioned before, endogenous ceramide levels are mainly regulated by its *de novo* synthesis from sphinganine and acyl CoA, which is catalyzed in mammals by a family of 6 ceramide synthases (Lass family) (Pewzner-Jung et al., 2006; Teufel et al., 2009). These enzymes are encoded by different genes that are evolutionarily highly conserved among eukaryotes. They preferentially use different fatty acyl CoA substrates containing fatty acid chains of different length, thereby producing ceramides with different acyl chains (Lahiri et al., 2005; Mizutani et al., 2005; Spassieva et al., 2006; Laviad et al., 2008; Riebeling et al., 2003; Mizutani et al., 2006) . According to the key roles ceramide plays in physiological cellular events, ceramide synthase activity must be strictly regulated.

Research of the past decades has led to the identification and characterization of several genes implicated in lipid and sphingolipid metabolism. This therefore considerably enhanced the understanding of molecular mechanisms controlling lipid homeostasis in response to physiological and energy supply requirements.

In animals, the balance between lipogenesis, lipid storage and lipolysis is controlled according to the prevailing energy requirement (Hay and Sonnenberg, 2004; Zechner et al. 2005). In times of energy abundance, lipogenesis is induced, leading to the synthesis of triacyl glycerides, which are deposited in adipose tissues in mammals and in the fat body in dipterans (Canavoso et al., 2001; Rosen and Spiegelmann, 2006). On the contrary, in conditions of energy depletion, triacyl glycerides are hydrolyzed to make fatty acids and diacylglycerols available for energy supply (Arrese and Wells, 1997; Gibbons et al. 2000; Patel et al., 2005). Chronic unbalancing of these three homeostasis pathways is involved in the pathology of obesity, insulin resistance and lipo dystrophy syndromes in humans (Khan et al., 2006; Simba and Garg, 2006).

Despite great progresses and a huge amount of knowledge registered in lipid research, the roles of the genes implicated in the regulation of lipid and sphingolipid metabolism are far from being completely elucidated, due to the unavailability of viable knockout animals for these genes, which mostly show prenatal lethality. A gene implicated in Drosphila larvae development was recently discovered (Bauer et al, 2009). This gene named *schlank* encodes a Drosophila homologue of Lass/ceramide synthases family, controlling the de novo synthesis of ceramide. In addition to controlling the *de novo* ceramide synthesis, *schlank* plays a crucial role in regulating lipid homeostasis in fly by maintaining the balance between lipogenesis and lipolysis. On the one hand, *schlank* activate the Sterol Regulatory Element Binding Protein (SREBP), which induces the fatty acid synthase and acyl-CoA carboxylase genes to synthetize palmitate. On the other hand, *schlank* controls triacylglyceride mobllilisation by limiting   the   activity   of   the   triacylglycerol   lipase.

4 Aim of the study

The first aim of this work is to generate mammalian cells continuously expressing a long hairpin RNA (lhRNA) of 800 bp targeting the GCS- and the Psap-mRNA for RNAi dependent sequence specific degradation respectively and therefore down regulating the expression of the corresponding genes. To achieve this purpose, we engeneered plasmids encoding for an inverted repeat of 1600 bps homologous to GCS- and pSap-mRNA respectively, which was placed under the control of a RNA polymerase II promotor. These plasmids were then transfected into mammalian cells, which then were subjected to analysis with regard to sphingolipid depletion and storage respectively.

GCS catatysed the transfer of glucose from UDP-glucose onto Ceramide to generate GlcCer, the core unit of higher membrane glycosphingolipids, und pSap, through its derivatives –the sphingolipid activator proteins (SAPs A-D), which are small, non-enzymatic glycoproteins – stimulates the lysosmal degradation of various sphingolipids.

With this work we wanted to prove the suitability of endogenously plasmide synthetised lhRNAs for inducing permanent knockdown of genes in mammalian cells. The choice of the glucosylceramide synthase (GCS) and the prosaposine (pSap) genes as our silencing targets finds its explanation in the existence of cells, which are deficient for these genes, therefore deficient in glycosphingolipids for the GCS deficient and accumulating glycosphingolipids for the pSap deficient cells. This allows an adequate follow up of the phenotype of the lhRNA cells by comparison with that of the deficient cells.

The second aim of this work is to analyse the involvement of the *Drosophila schlank* gene in ceramide synthesis and in the regulation of fat body storage and mobilisation in S2 cells using RNAi.

5   Results

5.1 GCS-and pSap-RNAi in Mammalian cells

5.1.1 BLAST Research

The first important step for this work was the knowledge of the glycosylceramide synthase (GCS) and the prosaposine (Psap) cDNA sequences. This knowledge was necessary for designing appropriated primers to be used in generating the 800 Bps cDNA fragments and subsequently the respective 1600 Bps inverted repeats in pcDNA3.1. Effectively, BLAST research (NCBI Genbank, Locuslink, Proteome) was performed and the cDNA and sequences of interest were identified. These sequences are given in the Appendix (pages 85-89).

Table 1 shows the homologies estimated in percentage (%) of the human GCS and pSAP proteins to those of mouse, rate, *D. melanogaster* and *C. elegans*.

Proteins	Species			
	Mus musculus	Rattus norvegicus	Drosophila melanogaster	C.elegans
GCS	98%	97%	53%	45%
pSAP	64%	66%	25%	29%

Tabelle 5-1: Homology between the human GCS and Psap protein sequences with those of Mus musculus (Mm), Rattus norvegicus (Rn), *Drosophila melanogaster (Dm)* and *C. elegans.*

Based on the open reading frame (ORF) of the identified GCS and prosaposine-gene sequences, appropriated primers were designed and used to generate the cDNAs mentioned above (table 5-2)

Nr	mRNA Name	Orientation-	Sequence (5'-3')
O131	GCS	f	GAG T*GG ATCC* AG ATG GCG CTG CTG GAC CTG GCC TTG GAG *BamHI*
O132	GCS	f1	GAG TTA *CTC GAG* ATG GCG CTG CTG GAC CTG GCC TTG GAG *XhoI*
O134	GCS	r	GAG T*GA ATT C* GA AAC TGA GAA ATT GAA TAT GAG CCA G *EcoRI*
O140	SAP	f	CGA C*GG ATC C*AG ATG TAC GCC CTC TTC CTC CTG GCC AGC *BamHI*
O141	SAP	f1	CGA C*CT CGA G*AG ATG TAC GCC CTC TTC CTC CTG GCC AGC *XhoI*
O142	SAP	r1	GTA C*GA ATT C*AG CGC ACA GAT CTC CTT GGG TTG CAT G *Eco RI*

Table 5-2: primer sequences
This table contains the primers used to design the respective cDNA fragments of 800 Bps destined to generate the GCS- and Psap-inverted repeats in pcDNA3.1. The numbering is in accordance with the consecutive numbering in the internal Science Lab Database (SDLB). Depending on their orientation with regard to the respective sense and anti sense strands of cDNA fragments, the primers are denoted as: f (forward = sense strang), f1 (f with another restriction site). Recognition sites for endonucleases to generate restriction ends appropriated for sub cloning, are underlined.

5.1.2 Generation of psap-RNAi and GCS-RNAi cell lines

Firgure 5-1 depicts a summary of the strategy used here to generate the lhRNA and the RNAi cell lines.

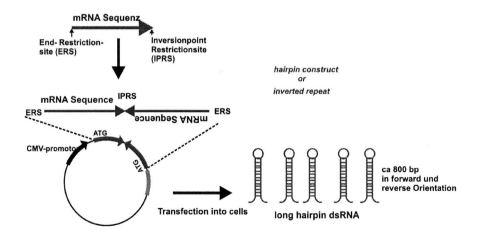

Figure 5-1. Summary of the strategy used to generate the lhRNA cell lines

5.1.2.1 Generation of the 800bp forvard and reverse cDNA fragments

To generate the 800 Bps forward and reverse cDNA fragments, the primers listed in Table 5-2 were used. While the 800bp fragments of pSap-cDNA were generated by PCR (s. Materials and Methods) using a prosaposin-cDNA containing clone availbable in our lab, those of GCS-cDNA were generated by RT-PCR using total RNA isolated from HeLa cells (s. Materials and Methods). The cDNA fragments were amplified by PCR and were then inserted into pBlueSKript vector to generate clones.

5.1.2.2 Generation of the pcDNA3.1 expressing the inverted repeat

The plasmids (pcDNA3.1) expressing the inverted repeats were generated in different ways. The pcDNA3.1-GCS-inverted repeat was generated by direct insertion of the 800bp forward and reverse cDNA fragments digested with BamHI/EcoRI and XhoI/EcoRI respectively into the BamHI/XhoI site of pcDNA3.1. (Fig.5-2 A). Instead, the pSAP-inverted repeat in pcDNA3.1 was generated by performing a double insertion of the same 800 Bps pSAP-cDNA fragment digested with XhoI/EcoRI into the dephosphorylated XhoI site of pcDNA3.1 (Fig.5-2 B). It is to note that the success rate in generating these clones was very low, which is due to the capacity of the inverted repeat to build a secondary structure that is completely or at least in part recombined and eliminated by bacterial recombinases during the replication process. Fig.5-2C and D shows the GCS- and the pSAP-inverted repeats in pcDNA3.1 after enzymatic digestion followed by agarose gel electrophoresis respectively.

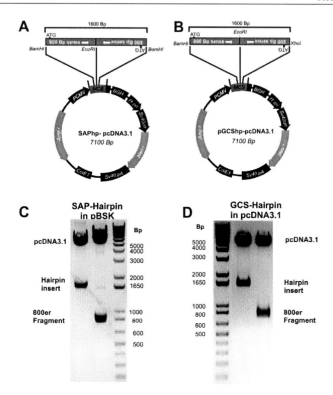

FIG. 5-2. Generation of the GCS- and pSAP-hairpin expressing pcDNA3.1 vectors. The GCS-hairpin-pcDNA3.1 was generated by inserting two GCS-cDNA fragments of 800 bp digested by BamHI/EcoRI and XhoI/EcoRI respectively into the pcDNA3.1 digested with BamHI/EcoRI (A). The Psap-hairpin-pcDNA3.1 was generated by performing a forward reverse insertion of a Psap-cDNA fragment of 800 bps, digested by XhoI/EcoRI into the pcDNA.3.1 clone digested by XhoI and then dephosphorylated (B). C. Enzymatic digestion of the GCS-hairpin-pcDNA3.1 (lane1: 1KB ladder, lane1: BamHI/XhoI, lane3: BamHI/XhoI/EcoRI). D. Enzymatic digestion of the pSAP-hairpin-pcDNA3.1 clone (lane4: XhoI, lane5: XhoI/EcoRI lane6: 1KB ladder)

5.1.2.3 Generation of the RNAi cell lines

After their successful cloning, the GCS-hairpin expressing plasmid was transfected into HeLa cells, human fibroblasts and into mouse melanocytes, while the pSAP-hairpin expressing plasmid was transfected into HeLa cells and mouse fibroblasts. The inverted repeats are transcribed under the control of the Polymerase II promoter. Following the transfection, cells were maintained under selection of G418 (300 –1200 µg/ml) for 30-40 days and were then analysed.

5.1.3 Molecular analysis of the RNAi cell lines

5.1.3.1 RT-PCR and Nothernblot analysis

The pSap- and the GCS-RNAi HeLa cells had been exclusively subjected to RT-PCR and Nothernblot to analyse the RNAi knockdown effect on pSAP and GCS due to the lhRNA expression. Total RNA was isolated from controls and RNAi cells, and RT-PCR performed using appropriate primers. Effectively, in contrast to the respective controls, no GCS- or Psap-mRNA has been detected in the GCS- and pSap-RNAi HeLa cells, therefore providing the first line of evidence that an RNAi induced degradation of the GCS- and pSap-mRNA has occurred in the respective RNAi cells (Figure 5-3) (Diallo et al., 2003)

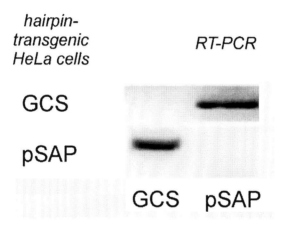

*hairpin-
transgenic
HeLa cells*   *RT-PCR*

GCS

pSAP

GCS   pSAP

Figure 5-3. RT-PCR analysis. Total RNA was isolated from GCS- and Psap-RNAi HeLa cells and subjected to RT-PCR as indicated in materials and methods. Each cell line was used as a control for the other.

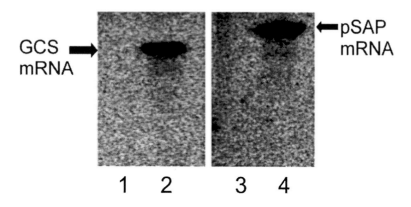

Fig. 5-4. Nothernblot analysis. Total RNA was extracted from Psap-, GCS-RNAi and normal control HeLa cells, and Nothernblot was performed, using ^{32}P-labeled cDNA of pSAP (lanes 1 and 2) and GCS (lanes 3 and 4) respectively. In contrast to normal controls, no mRNA could be detected in pSAP-(lane2) and GCS-RNAi (lane 4) HeLa cells.

To support the results obtained from RT-PCR, total RNA was isolated from normal and RNAi cells, blotted onto a nylon membrane, and probed against ^{32}P-labeled human GCS- and pSap-cDNA respectively. Similarly to RT-PCR, Nothernblot analysis detected mature GCS- and pSap-mRNA in the normal controls, but not in the GCS and pSap-RNAi cells (Fig.5-4).

It is known that foreign lhRNA, when it is taken up by mammalian cells, triggers the interferon cascade, leading to apoptotic cell death. A highly elevated oligoadenylat synthase (OAS) level in cells is a characteristic signal of an induction of the interferon response. To assess whether or not the endogenously synthetized lhRNAs trigger the interferon pathway in the respective RNAi HeLa cells, Nothernblot analysis was performed using an appropriate radiolabeled probe directed against OAS-mRNA to detect an eventual upregulation of OAS activity in the GCS-RNAi HeLa cells. However, this analysis revealed no difference in the OAS mRNA level between normal controls and treated cells (Figure 5-5 B lines 6 and 7), indicating that endogenously generated lhRNA is not or less potent in triggering the interferon cascade in cells. On the other hand, the RNAi cells revealed no characteristic signs of apoptosis. To assess the integrity of the total RNA used for Nothernblot analysis, prosaposine mRNA was detected in parallel using an appropriated radiolabeled prosaposine derived cDNA probe (Figure 5-5 B, lines 4 and 5).

For lhRNA to trigger RNAi, it must be processed by Dicer to siRNAs, which are the proper inducers of the sequence specific degradation of the target mRNA. Accordingly, our plasmid synthetised lhRNA is expected to be processed into siRNAs. Nothernblot analysis using an appropriate radio labeled GCS-cDNA probe effectively identified GCS-lhRNA derived siRNA in the GCS-RNAi HeLa cells, but not in the normal control (Figure 5-5 A, line 2 and 3)

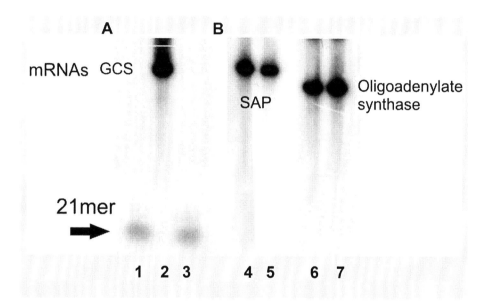

Figure 5-5. Nothernblot Detection of GCS derived siRNAs and OAS mRNA in control and GCS-RNAi HeLa cells by. Total RNA was extracted from GCS-RNAi and normal control HeLa cells, and Nothernblot was performed, using ^{32}P-labeled GCS- and OAS-cDNA. A. siRNA detection: line 1 standard siRNA; line 2: normal control; line 3: GCS-RNAi HeLa cells. B. OAS detection: lines 4 and 5: control of RNA integrity by detecting psap in control and GCS-RNAi cells respectively; line 6: OAS in control cells; line 7: OAS in GCS-RNAi cells.

5.1.3.2 Westernblot analysis

The level of GCS protein in normal controls and GCS-RNAi cells was determined by Westernblot analysis using a rabbit anti-GCS antibody. Figure 5-6 shows the results as obtained after autoradiographic detection. In contrast to the respective normal controls, no GCS protein could be detected in the GCS-RNAi cells, thus supporting the results obtained by RT-PCR and Nothernblot analysis. In mouse melanocytes, two forms of the GCS proteins are detected, the mature and the immature forms.

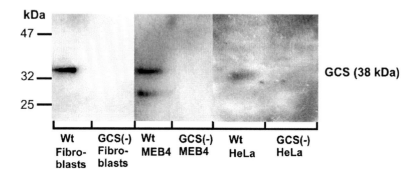

Fig. 5-6. Westernblot analysis. To analyse the level of GCS protein, total protein was obtained from wild type HeLa cells, human primary fibroblasts, and mouse melanocytes, and from the respective respective GCS-RNAi cells. The proteins were separated by SDS-PAGE and blotted onto a PVDF membrane. Immunodetection was performed using a rabbit anti-GCS antibody as a primary antibody and a horseradish peroxidase (HRP)-coupled antirabbit-IgG as a secondary antibody. GCS was detected by the HRP-activated chemiluminescence substrate using Lumigo.

5.1.3.3 Immunnocytochemical analysis

In contrast to GCS protein, the pSap level in the pSap-RNAi HeLa cells was determined by immunocytochemical analysis unsing a goat anti-Sap-D antibody as primary antibody and an HRP-linked antirabbit goat IgG as secondary antibody (Materials and Methods). As a cytosolic control, ß-adaptin was detected using an anti-ß-adaptin antibody. In this analysis, both Psap and ß-actin could be detected in normal control cells, while only ß-actin was detectable in pSap-RNAi HeLa cells (Fig.5-7)

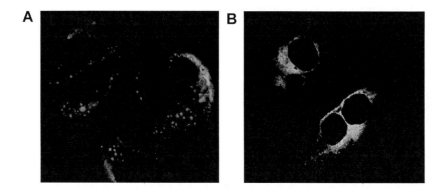

Figure 5-7. Immunofluorescence detection of prosaposine. Normal control and pSap-RNAi HeLa cells plated on cover slips were fixed and incubated with SAP-D antibody and the ß-adaptin antibody as cytosolic control. (A) Wild type cells showing strong fluorescence signal at their reticular structure. (B) No fluorescence could be detected in the pSap-RNAi cells.

5.1.3.4 GCS-RNAi MEB4 cells lost the ability to process melanin

Another visual assay to analyse GCS protein expression is based on the pigmentation of MEB4 cells. Van Meer's group (Sprong et al., 2001) reported that in a GCS knockout mouse melanoma cell line, tyrosinase was mislocated or retained within the Golgi apparatus, whereas in wild type cells, it was usually transported within specialized vesicles to the melanosomes. In the melanosomes, tyrosinase converts tyrosine to L-DOPA, which is the first rate-limiting step in the synthesis of melanine responsible for pigmentation. Inhibition of this step completely blocks the synthesis of melanin and pigments. Sprong et al. (2001) also showed that other melanosome-specific proteins reached their final destination, suggesting that tyrosinase is transported within vesicles enriched in glucosylceramide (GlcCer), whose synthesis is catalysed by GCS (glucosylceramide synthase). To investigate the effect of the RNAi dependent silencing of the GCS on melanine and pigment production in mouse melanoma (MEB4) cells, wild type and RNAi cells were harvested and peletted by centrifugation. Effectively, as reported by Van Meer's group, the RNAi MEB4 cells lost their pigmentation and became white, while the normal control cells still conserved their pigmentation and remained black (Figure 5-8.)

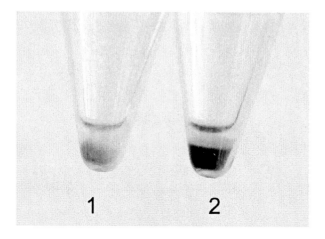

Fig. 5-8. GCS-RNAi MEB4 cells lose the capacity to process melanine. After removing medium, controls (1) and GCS-RNAi MEB4 (2) cells were washed two times with PBS, harvested and pelleted by centrifugation. The deficiency of glucosylceramide in MEB4 cells disturbs their capacity to transport tyrosinase to the melanosomes, whre it is supposed to process melanin. Consequently, these cells lose their pigmentation, turning from dark brown (1) to almost white (2). 1. wild type, 2 GCS-RNAi MEB4 cells.

5.1.4. Sphingolipids analysis of GCS deficients cells

5.1.4.1 Metabolic labeling with [^{14}C]-serine

Down regulation of the glucosylceramide synthase activity ultimately leads to a reduction of the glycosphingolipid levels in the cells. To analyze the effect of the RNAi mediated down regulation of the GCS on the sphingo- and glycosphingolipid levels, GCS-RNAi and the normal HeLa cells were metabolically labelled with [^{14}C]-serine, their sphingolipid extracted and subjected to TLC analysis. As another control, PDMP treated HeLa cells were as well metabolically radiolabeled and analyzed in parallel. The first observation from this analysis is that the most representative sphingolipids in HeLa cells are ceramide, globosides and sphingomyelin. Glucosylceramide, lactosylceramide and gangliosides represent minor components, whereby glucosylceramide, lactosylceramide and globosides appear as triple bands due to their heterogenous ceramide composition. Fig.5-9 shows the patterns of glycsphingolipids as obtained after TLC analysis. As compared to normal controls, GCS-RNAi and PDMP treated HeLa cells contained significantly reduced levels of radiolabelled glycosphingolipids. Fig.5-10A shows another TLC analysis of the neutral fraction of the GSLs in chloroform/ methanol/ water, 70/25/5, V/V/V with subsequent quantification of the incorporated radioactivity (Fig.5-10B). In contrast to reduced levels of glucosylceramide,

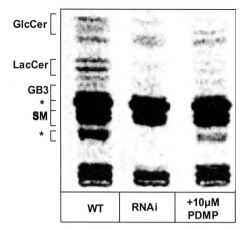

Fig. 5-9. GCS-RNAi, PDMP treated and normal controls HeLa cells were incubated with [^{14}C]serine (1μCi/ml) for 24h. Then the medium was exchanged and the cells were chased with a medium containing unalabeled serine for 120h. After harvesting of the cells, the neutral fraction of the glycosphingolipids was isolated, equal amounts of proteins were separated by TLC in chloroform/ methanol/ CaCl2 (0,22%), (V/V/V). The radioactive spots were identified by autoradiography as described in Materials and Methods. Abbreviations used: GalCer, galactosylceramide, Glccer, glucosylceramide, LacCer, lactosylceramide, RNAi, RNA interference, SM, sphingomyelin, WT, wild type, *, non identified bands .

lactosylceramide and globosides, the GCS-RNAi HeLa cells show elevated levels of ceramide and sphingomyelines as compared to the normal controls (Fig.5-10A-B). This is consistent with the down regulation of the glucosylceramide synthesis by the GCS that is thought to be a regulator of the ceramide level in cells.

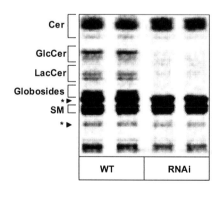

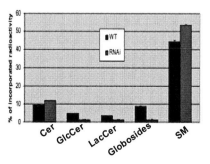

A   B

Fig. 5-10. TLC analysis of [¹⁴C] serine labeleld neutral GSLs from normal control and GCS-RNAi HeLa cells. A. GCS-RNAi and normal controls HeLa cells were incubated with [¹⁴C]serine (1μCi/ml) for 24h. Then the medium was exchanged and the cells were chased with a medium containing unlabeled serine for 120h. After harvesting of the cells, the neutral fraction of the glycosphingolipids was isolated, equal amounts of proteins were separated by TLC, and the radioactive spots were identified by autoradiography as described in Materials and Methods. **B.** Quantitation of the GSLs evaluated in percentage of incorporated radioactivity. Abbreviations used: Cer, ceramaide, Glccer, glucosylceramide, LacCer, lactosylceramide, RNAi, RNA interference, SM, sphingomyelin, WT, wild type.

Further, as compared to wild type, GCS-RNAi HeLa cells show significantly reduced levels of the acidic glycosphingolipids GM1, GM2 and GM3 and elevated levels of sulfatides (Fig.5-11).

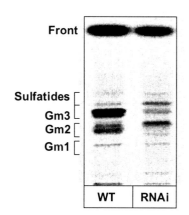

Front

Sulfatides

Gm3
Gm2
Gm1

WT   RNAi

Fig.5-11. GM3 and GM1 levels are decreased and sulfatides levels increased in GCS-RNAi HeLa cells. GCS-RNAi and normal HeLa cells were incubated with [¹⁴C]serine (1μCi/ml) for 24h. Then the medium was exchanged and the cells were chased with a medium containing unlabeled serine for 120h. After harvesting of the cells, the acidic fraction of the glycosphingolipids was isolated, equal amounts of protein were separated by TLC and the radioactive spots identified by autoradiography as described in Materials and Methods.

Similarly, TLC analysis of metabolically radio labeled sphingolipids from GCS-RNAi mouse melanocyte cells show them to incorporate less radioactivity into glucosylceramide, lactosylceramide, and globosides as compared to those from normal controls (Fig.5-12). On the opposite side, they show elevated levels of galactosylceramide and sphingomyelin (Fi.5-12), indicating that the lack of GCS activity lead to increased ceramide incorporation into these compounds.

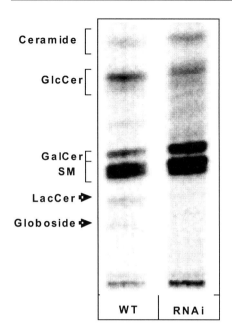

Ceramide

GlcCer

GalCer
SM

LacCer ▷

Globoside ▷

WT | RNAi

Fi.5-12. Metabolic labeling with [^{14}C]-serine. GCS-RNAi and normal controls MEB4 cells were incubated with (1µCi/ml) for 24h. Then the medium was exchanged and the cells were chased with a medium containing unlabeled serine for 120h After harvesting of the cells, the neutral fraction of the neutral glycosphingolipids was isolated, equal amounts of protein were separated by TLC on a borate plate and the radioactive spots were identified by autoradiography as described in Materials and Methods. Abbreviations used: Cer, ceramaide, GalCer, galactosylceramide, Glccer, glucosylceramide, LacCer, lactosylceramide, RNAi, RNA interference, SM, sphingomyelin, WT, wild type.

5.1.4.2 Metabolic labeling with [^{14}C]-galactose

To further investigate the RNAi mediated silencing of GCS, normal controls, GCS-RNAi and PDMP (10µM) treated HeLa cells were metabolically labeled with [^{14}C] galactose and their sphingolipids were analysed by TLC. In cells, galactose is transformed by epimerisation into glucose, which will be then transferred by GCS onto ceramide to build glucosylceramide. The patterns of the glycosphingolipids are given in Figure 5-12 as obtained after TLC and autoradiography. This analysis as well shows that the untreated cells incorporate significantly more radioactivity into glucosylceramide, lactosylceramide and globosides than the GCS-RNAi and the PDMP treated cells, indicating that GCS has been down regulated by PDMP as well as by RNAi. However, RNAi and PDMP seem to act differently. While RNAi acts on all three glucosylceramid bands, with a minor extent on the upper one, PDMP rather affected the first two upper bands and at a very low extent on the third one (Fig.5-12). Unidentified bands are represented by *. Some of these bands appearing below the front of the tlc in the RNAi and the PDMP treated cells but not in the normal controls, could be derivatives of galactose that by default of entering the glucosylceramide synthesis pathway, enters alternative metabolic pathways.

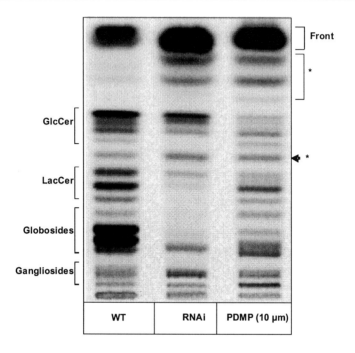

Fig.5-13. Metabolic labeling with [¹⁴C]-galactose. Normal controls, GCS-RNAi and PDMP (10μM) treated HeLa cells were incubated with [¹⁴C]-galactose (2μCi/ml) for 10h. Then the medium was removed, cells washed two times with PBS and harvested. Sphingolipids were then extracted and equal amounts of radioactivity separated by TLC and visualized by phosphoimaging. Unidentified bands are indicated by *. Abbreviations used: Cer, ceramide, Chol, cholesterol, GlcCer (glucosylceramide), LaCcer (lactosylceramide), PDMP, para-dimethylpyrolidone, WT, wild type.

5.1.5 Lipid analysis of prosaposine deficient cells

5.1.5.1 Metbolic labeling with [¹⁴C] serine

Since several decades, it is known that prosaposine deficient cells accumulate lysosomal sphingolipids, such as ceramide, glucosylceramide, lactosylceramide, globosides and gangliosides (Bradova et al. 1993; Harzer et al. 1989; Hulkova et al. 2001). Normal controls and pSAP-RNAi HeLa cells were radio labeled with [¹⁴C] serine, their sphingolipids extracted and subjected to TLC analysis. As a control, wild type and prosaposine deficient fibroblasts from Harzer's patients were as well radio labeled and analysed in parallel. The patterns of labeled glycosphingolipids as obtained after TLC are shown in Figure 5-13.

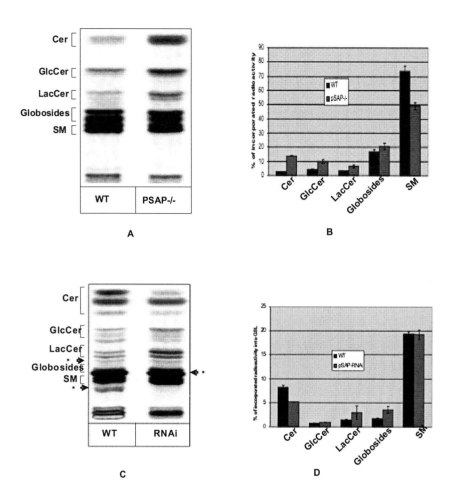

Fig.5-14. Psap-RNAi HeLa cells and Psap-/- human fibroblasts incorporate more radioactivity into sphingolipids than the corresponding normal controls. WT and Psap-/- human fibroblasts (**A, B**), WT and pSAP-RNAi HeLa cells (**C, D**) were incubated with [^{14}C]serine (1μCi/ml) for 24h. Then, the medium was exchanged and the cells were chased for 120h with a medium containing unlabeled serine. After harvesting of the cells, the neutral fraction of glycosphingolipids was isolated, equal amounts of proteins were separated by TLC and the radioactive spots identified by autoradiography and quantified as described in Materials and Methods. **A** and **B** show the patterns of GSLs from fibroblasts and their quanitation in percent (%) of incorporated radioactivity. **C** and **D** show the patterns of GSLs from HeLa cells and their quanitation in percent (%) of incorporated radioactivity. Unidentified bands are denotated by *. Abbreviations used: GlcCer (glucosylceramide), LacCer (lactosylceramide), SM (sphingomyelin), RNAi (RNA interference), WT (wilde type).

Compared to the respective normal controls, prosaposine deficient fibroblasts contained significantly increased levels of ceramide and glycosphingolipids. Similarly, prosaposine-RNAi HeLa cells contain highly elevated levels of lactosylceramide and globosides, moderately elevated level of glucosylceramide. However, in contrast to prosaposine deficient fibroblasts, the prosaposine-RNAi HeLa cells show decreased levels of ceramide compared to normal controls, which indeed could be explained when considering ceramide to be sequestred in the complexe glycosphingolipids that are not more hydrolysed. Interestingly, this decreased level of ceramide has been exclusively observed in early generations of prosaposine-RNAi HeLa cells.

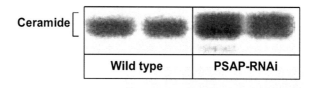

Fig.5-15. Incorporated radioactivity into ceramide from normal controls and pSAP–RNAi HeLa cells
Prosaposine-RNAi HeLa cells and normal controls were incubated with [^{14}C]serine ($1\,\mu$Ci/ml) for 24h. Then the medium was changed and the cells were chased with a medium containing unlabeled serine for 120h. After harvesting of the cells, the neutral fraction of the glycosphingolipids was isolated, equal amounts of proteins were separated by TLC and the radioactive spots identified by autoradiography as described in Materials and Methods
.

Metabolic labeling of late generations shows in contrast a highly increased level of ceramide compared to normal controls (Figure5-14), which was further confirmed by TLC analysis of of unlabeled ceramide.

Late generations of prosaposine-RNAi HeLa cells probably down regulate the overall biosynthesis of complexe glycosphingolipids, which consequently leads to ceramide accumulation over the time. Further, compared to wild type, prosaposine-RNAi cells show elevated levels of GM3, GM1 and sulfatides, and slightly reduced level of GM2 (Figure 5-15 A and B.)

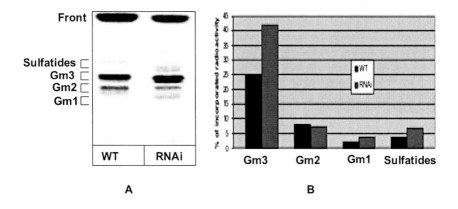

Fig.5-16. Psap-RNAi HeLa cells accumulate gangliosides and sulfatides. A. GCS-RNAi and normal controls HeLa cells were incubated with [^{14}C]serine (1μCi/ml) for 24h. Then the medium was exchanged and the cells were chased with a medium containing unalabeled serine for 120h. After harvesting of the cells, the acidic fraction of the glycosphingolipids was isolated, equal amounts of proteins were separated by TLC and the radioactive spots identified by autoradiography as described in Materials and Methods. **B.** Quantification of the glycosphingolipids evaluated as the percentage of incorporated radioactivity.

5.1.5.2 Metabolic labelling with 3H-sulfatide

In cells and tissues, in which sulfatide normally occurs, its degradation is catalyzed by arylsulfatase A (ASA) that needs the sulfatide activator Sap-B as an essential cofactor (Jatzkewitz and Stinshoff 1973). Mutations resulting in the loss of activity of matured SAP-B (Henseler et al. 1996; Kretz et al. 1990; Rafi et al. 1990; Zhang et al. 1990, and Holtschmidt et al. 1991) leads to metachromatic leucodystrophy (MLD) that is characterized by a lysosomal storage of sulfatides and other sphingolipids in different organs of the patient, particularly in the white matter of the brain and in the myelin of peripheral nerves.

We incubated wild type and pSap-RNAi HeLa cells with [^{3}H]-sulfatide to investigate their ability to catabolize it. Although the cell lines incorporated the same quantity of sulfatide, their hability to degrade it was different. While normal control cells degraded nearly 50%, the pSap-RNAi HeLa cells degraded only 20% of the incorporated sulfatide into galactosylceramide (Figure 5-17). In nerve and kidney cells sulfatide and galactosylceramide are naturally occurring substrates and are degraded into galactosylceramide, and into ceramide and galactose respectively. In HeLa cells, the degradation of sulfatide into galactosylceramide seems to occur, although to a minor extent. However, they do not seem to be able to further catabolize galactosyceramide into ceramide and galactose. Establishing the balance of the sulftide turnover, by adding the incorporated and unincorporated radioactivy, supported this result.

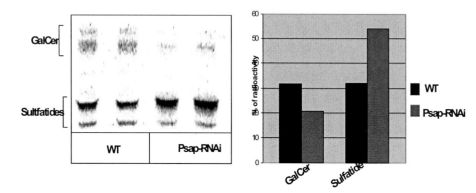

Figure 5-17. Sulfatide catabolism in control and pSap-RNAi cells. A. Control and pSap-RNAi cells were labeled with [³H]-sulfatide-BSA-complex and sphingolipids extracted and analyzed as described in materials and methods. **B.** Percentage of catabolized sulfatide.

5.1.5.3 Analysis of unlabeled lipids

TLC analysis of unlabeled sphingolipids from normal controls and late generations of prosaposine-RNAi HeLa cells was performed. The pattern of glycosphingolipids as obtained after TLC is given in Fig.5-18A. The prosaposine-RNAi HeLa cells show significantly increased levels of lactosylceramide, globosides and to minor extent glucosylceramide as compared to normal control cells. Fig.15-18B show a densitometric quantification of the glycosphingolipids evaluated in µg GSLs/mg of proteins.

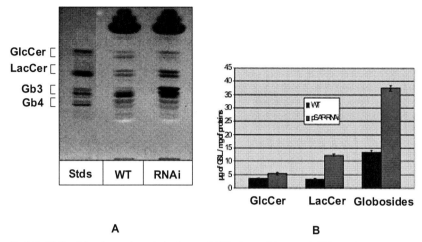

A   B

Fig.5-18. TLC and densitometric analysis of neutral GSLs from wild type and GCS-RNAi HeLa cells.
A. Unlabeled neutral glycosphingolipids were extracted from normal control and prosaposine-RNAi HeLa cells
and separated by TLC in chloroform/methanol/wasser: 70/30/5 (V/V/V). The TLC plate were then developed
with orcinol and heated for 5 min at 100°C. **B.** Densitometric qunatitation of the glycophingolipids given in
μg/mg of proteins. Abreviations used: GlcCer, glucosylceramide, GSL, glycosphingolipids, GB3,
globotriaosylceramide, GB4, globotetraosylceramide, LacCer, lactosylceramide, pSAP, prosaposine, RNAi,
RNA interference, Stds, standards, WT, wild type.

The pSap-RNAi cells as well accummulate ceramide and GM3. Fig.5-19 A and B and Figure
5-20 A and B show the patterns of these sphingolipids as obtained after TLC analysis, and
their densitometric quantification respectively.

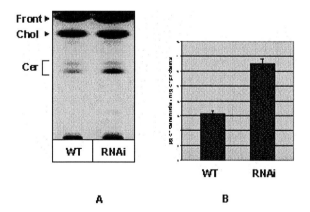

A   B

Fig. 5-19. TLC analysis of ceramide from normal controls and Psap-RNAi HeLa cells. A. Neutral GSLs
were isolated from WT and Psap-RNAi HeLa cells and separated by TLC in chloroform/methanol/acetic acid,
190/9/1 (V/V/V). To detect ceramide, the TLC plate was developed to saturation with a cupric sulphate solution

and heated at 180°C for 10 min. **B**. Densitometric quantification of ceramide from normal control and pSAP-RNAi HeLa cells given in µg ceramide/mg of total protein. Abbreviations: Cer, ceramide, Chol, cholesterol, RNAi, RNA interference.

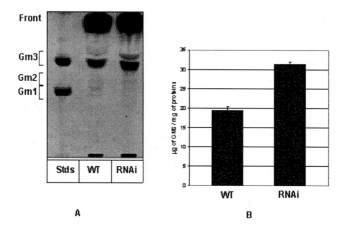

A   B

Figure 5-20. Psap-RNAi HeLa cells accumulate GM3. A. The acidic fraction of GSLs were isolated from WT and Psap-RNAi HeLa cells and separated by TLC in chloroform/methanol/CaCl2 (15 mM), 60/35/8 (V/V/V). To detect gangliosides the TLC plate was developed with a resorcinol solution and heated at 140°C for 10 min. **B.** Densitometric quantitation of GM3 from normal control and pSAP-RNAi HeLa cells given in µg GM3/mg of proteins. Abbreviations: Stds, standards, RNAi, RNA interference.

Further, unlabeled sphingolipids were extracted from wild type, prosaposine knockout and prosaposine-RNAi mouse fibroblasts and subjected to TLC analysis. The patterns of the sphingolipids obtained after TLC are shown in Figure 5-21. Similarly to the latter observation in HeLa cells and human fibroblasts, prosaposine deficient and prosaposine-RNAi mouse fibroblasts accumulate highly ceramide, glucosylceramide, lactosylceramide and globosides

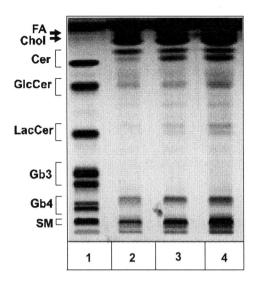

Fig. 5-21. pSAP knockout and Psap-RNAi mouse fibroblasts accumulate GSLs. Neutral glycosphingolipids from the respective cells were extracted and isolated. TLC separation was performed in chloroform/methanol/wasser 75/30/5 (V/V/V). The glycsphingolipids were visualized by staining with a copper sulphate solution and subsequent heating at 180°C. Lane1: standards, lane2: normal control, lane3: pSAP-KO, lane4: pSAP-RNAi. Abbreviations used: Chol, cholesterol, FA, fatty acids, GlcCer, glucosylceramide, , GB3 globotriaosylceramide, Gb4 globotriaosylceramide, LacCer, lactosylceramide, pSAP-KO, prosaposine knockout, SM, sphingomyelin.

5.1.5.4 Incubation with recombinant pSap abolishes sphingolipid storage in RNAi cells

Unlabeled neutral glycosphingolipids were extracted from prosaposine-RNAi HeLa cells incubated with recombinant prosaposin (0,2 µg/ml) for 96 h and subjected to TLC analysis in comparison to those from wild type and non treated prosaposine-RNAi cells. As expected, this prosaposine feeding drastically abolishes the storage of the neutral glycosphingolipids in the pSap-RNAi cells (Figure 5-22).

.

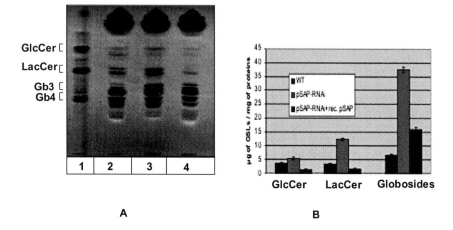

Figure. 5-22. Incubation with recombinant pSAP into the pSAP-RNAi HeLa cells abolishes GSLs accumulation. A. normal control cells and pSap-RNAi HeLa cells incubated or not with pSAP (0,2μg/ml) for 96h. After harvesting of the cells, the neutral fraction of glycosphingolipids were isolated and separated by HPTLC in chloroform/methanol/water 70/30/5 (V/V/V). The plate was developed with orcinol and heated for 5 min at 100°C. Lane 1: standards, lane 2: normal controls, lane 3: pSAP RNAi cells, lane 4: pSAP RNAi cells incubated with recombinant prosaposine (0,2 μg/ml). **B.** Densitometric quantitation of the GSLs giving in μg GSL/mg of proteins. Abreviations used: GlcCer, glucosylceramide, LacCer, lactosylceramide, GB3 globotriaosylceramide, Gb4 globotriaosylceramide, GSLs glycosphingolipids, RNAi, RNA interference)

The levels of glucosylceramide, lactosylceramide and globotriaosylceramide in the treated cells were reduced to 19%, 10% and 44,4 % down to levels of non treated pSap-RNAi cells.

Glucosylceramide and lactosylceramide were decreased even down to wild type levels

As well, the ceramide storage has been abolished by the prosaposine feeding, decreasing its level to 46% of the value of non treated pSap-RNAi cells and to the level of wild type cells (Figure 5-23).

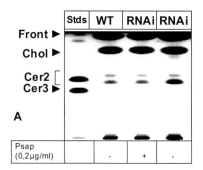

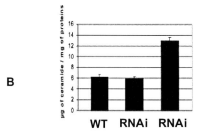

Fig.5-23. Incubating the Psap-RNAi HeLa cells with recombinant prosaposine abolishes ceramide storage. A.The neutral fraction of unlabeled GSLs was isolated from normal controls, Psap-RNAi and Psap-RNAi treated with recombinant Psap and separated by TLC in chloroform/methanol/acetic acid 190/9/1 (V/V/V). The plate was then sprayed with a cupric sulphate solution and heated for 10 min at 180°C. B. Densitometric quantification of ceramide evaluated in µg ceramide/mg of proteins

5.2 S2 Schlank-RNAi cells

5.2.1 Analysis of Schlank in larvae

5.2.1.1 Identification and molecular characterization of Schlank

In a search for genes controlling larval growth in *Drosophila*, two P element alleles of a novel locus, which was named *schlank*, were identified on the X-chromosome (Bauer *et al.*, 2009). Hemizygous *schlank*^{G0349} and *schlank*^{G0061} mutants are characterized by pronounced growth defects and larval lethality (Bauer *et al.*, 2009). Molecular analysis and genetic rescue experiments demonstrate that the lethality of the *schlank* alleles is linked to the *schlank* gene function. Both alleles are hypomorphic alleles and the transcription of the gene is strongly reduced.

Schlank is a protein containing six putative transmembrane (TM) domains, a Lag1 motif and a Hox domain, features that define it as a member of the among eukaryotes highly conserved Longevity assurance gene (Lass) family. As bona fide (dihydro) ceramide synthases (Mizutani et al., 2005; Venkataraman and Futerman, 2002), lass proteins catalyze the *de novo* synthesis of (dihydro)ceramide using sphinganine or sphingosine and fatty acyl-CoA with varying chain length as substrates. Their Lag1 motif, functionally required for (dihydro)ceramide synthesis, is contained within a stretch of 52 amino acids (Jiang et al., 1998). While the Lag1 motif of Schlank is located in the fourth TM domain (Bauer et al., 2009), its Hox domain, that is characteristic for homeodomain transcription factors, is located within the loop region in between the TM domains 1 and 2. Hox domains are found in only one distinct subclass of Lass proteins of arthropods and mammals, but not in yeast and plants (Venkataraman and Futerman, 2002). The function of Lass Hox domains is unknown.

5.2.1.2 Schlank regulates larval growth

Phenotypic analysis of hemizygous *schlank*^{G0349} and *schlank*^{G0061} mutants indicates an essential role of *schlank* in larval growth. Both mutants are significantly smaller compared to control animals (Figure 5-24 A, B and C). *schlank*^{G0349} mutants die as morphologically first instar larvae although they feed, as verified by feeding red coloured yeast. Animals carrying the weaker *schlank*^{G0061} allele also show retarded larval growth accompanied with variations in size (Figure 24 C). However, a fraction of the hemizygous mutants reach the third instar larval stage. Some of these animals even pupariate and give rise to adults which die shortly after eclosure with severe morphological defects.

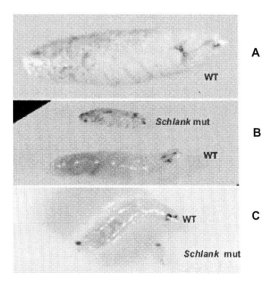

Fig.5-24. Phenotypic analysis of hemigygous $schlank^{G0349}$ **and** $schlankG0061$**.** Wild type larvae (**A**). Mutants are significantly smaller compared to normal control (**B**). $schlank^{G0349}$ mutants die as morphologically first instar larvae although they feed, as verified by feeding red coloured yeast. Animals carrying the weaker $schlank^{G0061}$ allele also show retarded larval growth accompanied with variations in size. A fraction of the hemizygous mutants reach the third instar larval stage (**C**). Some of these animals even pupariate and give rise to adults which die shortly after eclosure with severe morphological defects.

5.2.1.3 Lipid analysis in larvae

Qualitative and quantitative analysis of lipids in wild type, $schlank^{G0349}$, and in schlank over expressing larvae, revealed that the amount of ceramides (mainly Cer(d14:1/20:0) and Cer(d14:1/22:0), fatty acids (2.3-fold), as well as triacylglycerides (1.4-fold) were increased in larvae overexpressing *schlank* (*UAS*-schlank), whereas the ceramide (85% of the control) and triacylglyceride levels (88% of the control) were reduced in $schlank^{G0349}$ mutants (Bauer et al., 2009). Moreover, it has been found that the level of dihydroceramide was significantly decreased. In contrast, ceramide levels were upregulated upon overexpression of *schlank* in larvae using *hs*-Gal4 driver and UAS-*schlank*-HA effector lines (Bauer *et al.*, 2009)

5.2.2 Analysis of schlank in S2 cells

To support the results obtained from the analysis of the larvae and to ascertain that the difference observed in their lipid profiles is due to the influence of schlank on their respective endogenous lipid metabolism and not to an impaired lipid uptake from the yeast food, we transiently down regulated schlank expression in S2 cells by means of RNAi on one hand; on the other we generated schlank over expressing S2 cells and analysed their lipids.

5.2.2.1 S2 cells incubated with schlank-dsRNA are deficient in schlank-mRNA

Total RNA from normal control and schlank-dsRNA treated cells was prepared and RT-PCR performed to amplify the schlank-mRNA using appropriated primers. Interestingly, RT-PCR (Figure 5-25) analysis detected schlank-mRNA in the WT but not in the dsRNA treated S2 cells, suggesting, as expected, that the schlank-mRNA effectively has been targeted for degradation by the introduced homologous dsRNA, leading to a down regulation of the schlank-protein. On the other hand, GCS-mRNA has been detected in the both total RNA samples by RT-PCR to assess their integrity.

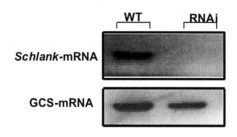

Fig.5-25 Schlank-dsRNA treated S2 cells are lacking of schlank-mRNA. Total RNA was isolated from normal control and schlank-dsRNA treated S2 and subjected to RT-PCR as indicated in materials and methods to amplify schlank-mRNA. As control, fly GCS-mRNA was amplified in parallel. In contrast to normal controls, no schlank-mRNA could be amplified in treated cells. GCS-mRNA was amplified in both treated and non treated cells.

5.2.2.2 Ceramide levels are reduced in Schlank-dsRNA treated S2 cells.

The first attempt to address the question whether schlank is really involved in dihydro(ceramide) synthesis in fly consisted of transiently down regulating its activity in S2 cells by means of RNAi and a subsequent analysis of their sphingolipids.

S2 cells were incubated 11 days with *schlank*-dsRNA as indicated in Materials and Methods to down regulate *schlank* expression by targeting its mRNA for sequence specific degaradation. Then during the last day (24h) of the dsRNA treatement, cells were metabolically labelled with $[^{14}C]$-serine (1μCi/ml). As a control, untreated cells were metabolically radio labelled and analysed in parallel. Untreated cells contained 15,3% and treated cells 15,6% of total radioactivity after the pulse and 4,9% and 5.3% of cell bound radioactivity after an additional chase with unlabeled serine for 120 h respectively. Respectively 8,8% and 9,5% from the latter were extracted to the total sphingolipid fraction.

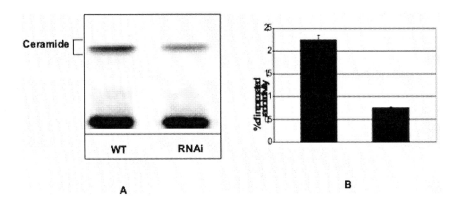

A

B

Fig.5-26. Ceramide level is decreased in schlank-RNAi cells. Normal control and schlank-dsRNA treated S2 cells (n=2) were metabolically labelled with [^{14}C]-serine, their sphingolipids extracted and separated by TLC in chloroform/methanol/acetic acid 190/9/1 (V/V/V) as indicated in materials and methods. **A.** TLC imaging. **B.** Quantitation der incorporated radioactivity into ceramide

To highlight the impact of dsRNA treatment on dihydroceramide synthesis in the S2 cells, TLC analysis was performed in chloroform/methanol/acetic acid 190/9/1 (V/V/V) (Figure 5-26 A-B). This analysis shows that the level of ceramide is effectively reduced in the dsRNA treated cells, as compared to normal controls. On the other hand TLC analysis of unlabeled sphingolipids from normal controls and dsRNA treated S2 cells confirms ceramide reduction in the latter (Figure 5-27 A-B). These results provide the first line of evidence supporting the idea of schlank to be involved in ceramide synthesis in Drosophila S2 cells.

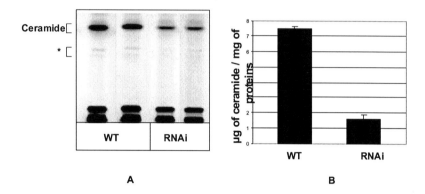

Fig.5-27. TLC analysis of cold lipids. A. Unlabeled sphingolipids were extracted from normal control and schlank-dsRNA treated S2 cells (n=2) and separated by 'TLC in chloroform/methanol/acetic acid 190/9/1 (V/V/V) as indicated in materials and methods. B. Densitometric quantitation of ceramide evaluated in μg ceramide / mg of proteins. Unidentified bands are denoted by *

Further, LC/APCI-MS analysis as well demonstrated that ceramide and glucosylceramide levels in schlank-dsRNA treated cells are reduced to at least half the amounts of wild type levels (Figure 5-28 A). This analysis shows that the down regulation of the main ceramide specie expressed in Drosophila (Cer(d14:1/20:0) and Cer(d14:1/22:0) lead to an increased incorporation of lower molecular ceramide species into glucosylceramide leading to formation of low molecular glycosphingolipid species (Figure 5-28 B)

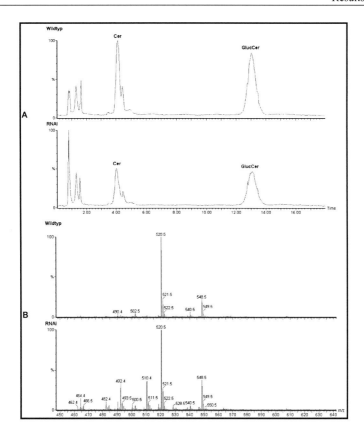

Fig.5-28. LC/APCI-MS analysis. Sphingolipid were isolated from normal control and schlank-RNAi S2 cells and subjected to LC/APCI-MS analysis as indicated in Materials and Methods. Significant downregulation of ceramides und glucosylceramide in schlank RNAi cells (**A**, LC profile see upper two panels. Cer(d14:1/20:0) and Cer(d14:1/22:0) are the most abundant species (**B**)

5.2.2.3 Schlank over expression in S2 ceramide results in highly elevated ceramide level.

Another strategy used to provide evidence for *schlanks* implication in ceramide synthesis in fly consisted in the generation of S2 cells over expressing the schlank protein. *Schlank* is assumed to be a proper dihydroceramide synthase or at least an essential component of this enzyme, and therefore, its over expression in cells should ultimately implicate an elevation of dihydro(ceramide) levels. To address this issue, schlank overexpressing S2 cells were generated, metabolically labelled with [14C]-serine and sphingolipids isolated and analysed by TLC in comparison to wild type and to *schlank*-dsRNA treated S2 cells as indicated above.

Figure 5-29 shows the ceramide pathern of ceramide as obtained after after TLC. As expected, the S2 cells over expressing the schlank gene synthesizes more ceramde than the normal controls and the schlank-dsRNA treated S2 cells, which on its side shows reduced ceramide level as compared to the normal controls (Figure 5-29 A, B)

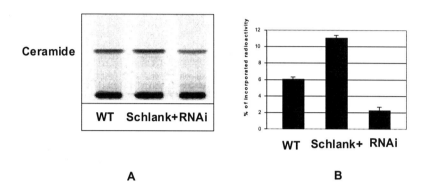

Fig.5-29. Ceramide level is elevated in schlank over expressing S2. A. normal control, schlank-RNAi and schlank overexpressing S2 were metabolically labelled with [^{14}C]-serine, their sphingolipids extracted and equal amounts of proteins separated by TLC in Chloroform/Methanol/Ac. Ac 190/9/1 (V/V/V). **B.** Quantitation of ceramide evaluated in percent (%) of radioactivity incorporated into ceramide.

5.2.2.4 Ceramide synthase essay in cells and larvae

(Dihydro)ceramide synthase assays and biosynthetic labeling studies using lysates from larvae (Bauer et al., 2009) or *schlank*-RNAi S2 cells provided further evidence that *Schlank* is required for (dihydro)ceramide synthesis. Hier we used lysates from normal controls and dsRNA treated S2 cells to perform the essay by incubating these lysates with radio labeled sphinganine ([^{3}H]sa) and palmitic acid as indicated in materials and methods. Then sphingolipids were extracted and equal amounts of protein separated on TLC and visualized by autoradiography (Figure 5-30 A-B). Expectedly, the lysate from the *schlank*-dsRNA treated cells synthesize less ceramide as compared to the lysate from the normal controls, which consistent with the idea of schlank to be dehydroceramide synthase.

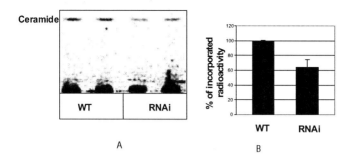

Fig.5-30 Ceramide synthase essay. Was performed as indicated in materials and methods. The dried lipids were resuspended in 40 μl of chloroform/methanol 1:1, V/V and separated on Silica Gel60 HPTLC plates (Merck). **A.** The separated lipid bands were visualized with a phosphoimager (FUJIX Bio Imaging Analyser 1000, Fuji Photo Film, Tokyo, Japan) and quantified by TINA 2.08 software (Raytest, Straubenhardt, Germany). **B.** % of incorporated radioactivity into ceramide

5.2.2.5 Schlank is a dihydroceramide synthase

The structural similarities between *schlank* and the members of the Lass proteins family and the observed correlation between its dowregulation and the decrease of ceramide and glycosphingolipid levels in the S2 cells clearly indicates that it is involved in dihydro(ceramide) synthesis. Now the question to be addressed remains whether *schlanck* is a dihydroceramide synthase using fattyacyl-coA and sphingoid bases or a ceramide synthase, which transforms dihydroceramide to ceramide. To answer this question, normal control and *schlank*-dsRNA S2 cells were metabolically labelled with $[^{14}C]$serine or $[^{14}C]$-acetate. The sphingolipids were extracted and subjected to TLC on a borate plate in chloroform/methanol 9/1 (V/V). This procedure allows the separation of ceramide from dihydroceramide. Figure 5-31 shows the ceramide pattern as obtained after the TLC. Intrestingly, the dihydroceramide level reveals to be higher in the wild type as compared to *schlank*-RNAi cells, which supports the idea that schlank is rather critical could be observed in the *schlank*-RNAi cells; rather, it is metabolised to the same extent to ceramide, as in the non treated cells. However, as compared to normal controls, the dihydroceramide level is reduced in the *schlank*-dsRNA treated cells. That indicates that *schlank* is rather implicated in the activity of the N-Acyl Sphinganine transferase leading to the formation of dihyrdroceramide (Figure 5-31 A and B). Furthermore, dihydroceramide is nearly completely transformed into ceramide in the RNAi cells, indicating that schlank is not involved in the step leading from dihydroceramide to ceramide (Figure 5-31 A and B).

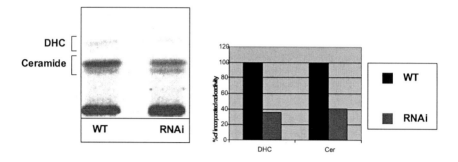

Fig.5-31 Schlank is a dihydroceramide synthase. A. Normal control and schlank-dsRNA treated cells (n=2) were metabolically labelled with [^{14}C]-serine. The sphingolipids were extracted and separated by TLC on a borate plate in Chloroform/Methanol 9/1 (V/V) to separate ceramide from dihydroceramide. **B.** % of radioactivity incorporated into ceramide and dihydroceramide.

5.2.2.6 Schlank regulates TAG lipases in *Drosophila*

Analysing the *Schlank* mutant larvae at RNA level revealed a significant up regulation of the *brummer* and *lip3* genes, encoding for an ATGL lipase (Grönke et al., 2005) and for a TAG lipase (Zinke et al., 1999; 2002) respectively. This finding is consistent with the idea that *schlank* controles lipolysis in Drosphila by inhibiting the TAG lipase, and is supported further by the observation that the expression of both lipases was decreased upon ubiquitous expression of *Schlank* in larvae using hs-Gal4:UAS-*schlank* transgenic flies ((Bauer et al. 2009). On our side, we performed a TLC analysis of the total lipid fraction from wild type and schlank-dsRNA treated S2 cells to support these results. Figure 5-32 A-D shows the results of this analysis as obtained by TLC. Effectively, as compared to the normal controls, the level of the faty acids is drastically elevated while that of the triacylglycerides is reduced in the *schlank*-dsRNA treated cells, which is consistent with an up regulation of the triacylglyceride lipase in the treated cells.

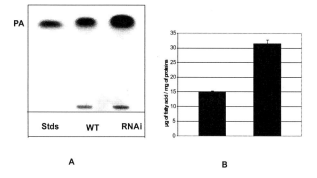

A

B

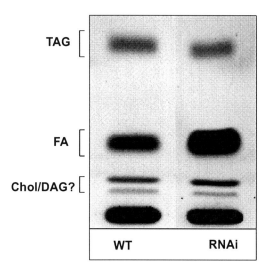

D

Fig.5-32. Schlank maintains the balance between lipolysis and lipogenesis. Unlabeled lipids were isolated from normal control and dsRNA treated S2 cells, separated by TLC and stained with cupric sulphate. **A.** Fatty acids. **B.** Densitometric quantitation of fatty acid evaluated in μg fatty acid / mg of proteins. **D.** TAG separation and visualisation.

6   Discussion

GCS- and pSap-RNAi mammalian cells. RNAi has developed to an important tool in functional analysis of genes in cellular and animal systems, and promises as well to develop to an efficient weapon in fighting infection diseases. Soon after its discovery, it has been used to successfully knockdown gene function in *C. elegans*, *Drosophila melanogaster*, *N. crassa*, and plants. The first successful application of RNAi to sequence specifically silence genes in mammalian systems has been the injection of dsRNA into mouse embryonic cells and into early mouse embryos (Paddison et al. 2002b; Svoboda et al. 2000; Svoboda et al. 2001; Wianny and Zernicka-Goetz 2000). Lately, a pioneer work of Tuschl and coworkers (Elbashir et al. 2001) should have been essential in revolutionizing RNAi application in somatic mammalian cells. Referring to previous studies that identified siRNAs, processed from dsRNA, as the proper mediators of RNAi, they made the remarkable finding that transfecting synthetic 21-nt siRNA duplexes into mammalian cells effectively and specifically silences endogenous genes without triggering unspecific interferon responses, or even if, to an undetectable level (Elbashir, 2001). Since then, different RNAi based techniques reaching from using synthetic siRNAs at porpurses of transient RNAi (Donze, 2002) to stabile, plasmid or virus triggered RNAi (Brummelkamp et al. 2002; Lee et al. 2002; Miyagishi and Taira 2002a; Paddison et al. 2002a; Paul et al. 2002; Sui et al. 2002) have been developed and are routinely used to knock down genes in mammalian systems. Plasmid synthetized siRNA or dsRNA to trigger stabile RNAi offers the advantage to considerably reduce the induction of dsRNA dependent apoptotic pathways and to produce long lasting down regulation effect, allowing better analysis of the resulting phenotype over longer periods of time. On the other hand, dsRNA reveals to achieve stronger RNAi effect than single siRNA, due to the fact that Dicer processing of the dsRNA generates a pool of variable siRNAs leading to a highly cooperative targeting of the mRNA of interest.

In this work, we developed an RNAi system aiming at the generation of mammalian cells durably expressing a dsRNA of 800 Bps targeting the glucosylceramide synthase (GCS) or the prosaposine (Psap) genes for silencing. This RNAi system is based upon an eukaryotic expression plasmide (pcDNA3.1) in which wir inserted an inverted repeat of 1600 Bps consisting of the stretch encolmpassing the first 800 Bps of GCS or pSap-Cdna respectively. The transcription of these invertead repeats is under the control of a RNA polymerase II promotor.

The analysis by RT-PCR, Notherblot, Westernblot or Immunocytochimie of the respective RNAi cells shows a strong reduction of the GCS- and Psap-mRNA and their corresponding GCS- and Psap-proteins, as compared with normal controls. Sphingo- and glycosphingiolipids analysis is in agreement with these results. Compared to normal control cells, GCS-RNAi cells show decreased levels of glucosylceramide and its derivatives, e.g. lactosylceramide, globosides and gangliosides on the one hand; on the other they accumulate ceramide and sphingomyeline. These findings are consistent with the down regulation of the glucosylceramide synthase. In contrast to GCS-RNAi cells, pSap-RNAi cells significantly accumulate ceramide, glucosylceramide, lactosylceramide, globosides and gangliosides. All these glycosphingolipids are as well known to accumulate in prosaposine deficient fibroblasts from Harzer's patients. These accumulated sphingo- and glycosphingolipids are normally degraded by saposin dependent lysosomal exohydrolases. This accumulation become particularly evident with unlabelled glycosphingolipids separated by tlc and visualised with specific staining solution, orcin for glucosylceramide, lactosylceramide and globotriaosylceramide; resorcine for GM3 and cupric sulfate for ceramide. Further, due to this glycosphingolipid storage, the psap-RNAi cells become more and more voluminous compared to the control cells. Incubating the pSAP-RNAi cells with recombinant pSap-protein drastically abolishs the lysosomal accumulation of sphingolipids.

On the other hand, the endogenously generated lhRNA don't seem to significantly trigger unspecific pathways, like the interferon cascade leading to apoptotic cell death. Possible explanations for this observation could be that, contrarily to exogenously originating dsRNA, endogenously generated dsRNA is rapidly processed into active siRNAs, so that it cannot accumulate in the cytosol, where it may interact with players of the interferon pathway to induce it. On the other hand exogenously administrated dsRNA could possibly interact with membrane receptors involved in the activation of the interferon pathway, what don't happen, if the dsRNA is endogenously generated. It could also be possible that only those cells can escape selection, which produce long hpRNA in levels too low to induce interferon response, but sufficient to induce the degradation of the target mRNA

We show here that endogenously, plasmid synthetized dsRNA is a valid technique that can be applied to down regulate gene function in mammalian cell culture systems. However despite the advantages offered by this method, namely stronger silencing effect when compared to single siRNA and strongly reduced induction of non-specific reactions when compared to exogenously originating dsRNA, it reveals to be very laborious and quite restrictive. That means, the generation of the plasmid with the inverted repeat is very difficult on the one hand; on the other, as mentioned above, in some cases, the processing of long dsRNA by Dicer could generate siRNAs, cross targeting other genes. With this regard, the choice of the target sequence has to be done with precaution by taking in account this restrictive aspect. This surely explains the reason why siRNA and endogenously generated shRNA are getting the upper hand in the perspectives of using RNAi at purposes of functional genomic studies and therapeutics. The lowered efficiency of the single siRNA can be compensated by generating a pool of plasmids respectively encoding for different siRNAs that have been judiciously selected to specifically and cooperatively target along the mRNA of interest.

Despite many successes achieved in elucidating the function of numerous genes and accordingly several biochemical processes in invertebrates and plants, and lately in mammalian systems, RNAi, cannot yet be acquitted of side effects related to its action. Like other techniques used at purposes of functional genomics and therapeutic settings – for example the use of pharmacological compounds -, the introduction of siRNA into, or expressing shRNA or dsRNA in mammalian systems still causes problems of specificity, dosage and deliverance; the later essentially in living organisms. That is, for RNAi to full fill its promises to become the prominent tool above all else in the hand of large scale genomic screening on the one hand, and on the other in fighting viral infection and other human diseases, requires to be conceived to perfectly mimic the endogenous RNAi pathway directed by internal micro-RNAs and to harmonize with it. That can be achieved by designing endogenously, plasmid or lentiviral expressed shRNA or lhRNA with optimized specificity and efficiency on the one hand, on the other by monitoring and controlling the level of expressed shRNA or lhRNA to avoid an over saturation of the RNAi machinery.

Schlank-dsRNA Drosophila S2 cells. In addition to the homology shown by the *schlank* protein to the highly among eukaryotes conserved Lass family members, the results provided by the sphingolipid analysis in the S2-RNAi cells and the schlank mutant larvae clearly supports the idea that *schlank* is a bona fide *Drosophila* dihydroceramide synthase, a key enzyme of the *de novo* biosynthesis of sphigolipids (Mizutani et al. 2005, 2006, Pewzner-Jung et al. 2006). Till jet six Lass/CerS family members have been identified in mice (Futerman & Riezman 2005), although gene knock outs at purpose of functional studies on the organismic level are not available yet.

Compared to wild type, the ceramide levels are strongly reduced in *schlank* mutant larvae and the S2-RNAi cells. This ceramide depletion consequently affects the sphingolipid levels in the eukaryotic plasma membrane and probably explains the growth phenotype shown by the *schlank* mutants and the deleterious effects that this may have on cellular membranes and ceramide-dependent signalling processes. Another interesting finding is the up regulation of lipase activities in the schlank mutant larvae and RNAi S2 cells, inducing a strong TAG degradation and devoiding the larvae of lipid stores. This phenomenon, which has not been observed for the identified Lass family members in yeast, mammals or plants, is caused by an up regulation of the neuroendocrine hormone Akh in the larval ring gland and a dramatic induction of Akh-dependent TAG-lipases in the fat body (Bauer et al., 2009).

While the TAG levels, like in the *schlank* mutant larvae, are drastically reduced in the S2-RNAi cells, the fatty acid levels are in contrast rather elevated in the latter, which is consistent with an up regulation of the TAG degrading Lipases. This also supports the idea that *schlank* negatively regulates the lipase activity. The reduced Fatty acid levels in the *schlank* mutant larvae is also consistent with the weakened SREBP dependent fatty acyl-CoA synthase activity, which is in normal conditions positively regulated by the *schlank* protein. This suggests that the reduced level of TAG in the mutant larvae might have a double origin, namely the reduced fatty acyl-CoA levels and the up regulated lipase activity.

De novo sphingolipid biosynthesis is dependent on the availability of long-chain saturated fatty acids, which participate in the initial rate-limiting reaction involving the condensation of a fatty acyl -CoA and serine (Adachi-Yamada et al 1999, Batheja et. al 2003), and in the conversion of (dihydro)sphingosine to (dihydro)ceramid which involves amino acylation with a long chain fatty acid at carbon 2 of sphingosine. The latter reaction is catalysed by the Lass/CerS family of (dihydro) ceramide synthases and requires the Lag1 motif (Spassieva et al. 2006). Accordingly, we propose that *Schlank* may act on the one hand as a "metabolic sensor" for fatty acid levels in the Akh-expressing cells. When food is plentiful, *Schlank* expression would repress the transcription of *akh* in the ring gland and of lipases in the peripheral fat body, allowing SREBP-dependent activation of enzymes of fatty acid biosynthesis (Bauer et al, 2009). On the other hand, upon fasting, *schlank* transcription would become repressed. This would allow higher levels of Akh expression in the ring gland, resulting in the induction of lipases in peripheral fat body to initiate the break down of TAG body fat stores, thereby eventually generating fatty acid supply for cellular sphingolipid biosynthesis and energy homeostasis. In summary, we propose that Schlank acts as a metabolic sensor in the neuroendocrine ring gland cells. The dual function of Schlank in (dihydro)ceramide synthesis and in the regulation of Akh-dependent TAG lipolysis provides a sensitive mechanism by which cellular sphingolipid biosynthesis can be linked to the systemic control of body fat mobilization and storage to allow coordinated energy homeostasis during cellular and organismal growth (Bauer et al., 2009).

7   Materials and methods

7.1. Molecularbiological Methods

7.1.1. Methods for bacterial culture and storage

7.1.1.1. Storage

For storage and opportune availability of usual bacteria, they will be disposed in form of stab culture. For this purpose, bacteria from a culture solution are inoculated into 1ml of LB-Medium containing 1,5 % agar, 100 µg/ml ampicillin or 12 µg/ml tetracycline or 25 µg/ml kanamycin and incubated at 37°c for 12h. This stab culture can be stored at ambient temperature, in the darken for a maximum of 12 months. For stocks to be storied for several years, they are disposed in a glycerol solution 20% and placed at -80°c.

7.1.1.2. Bacterial culture

Bacteria are cultivated in LB-medium containing or not an antibiotic (100 µg/ml ampicillin or 12 µg/ml tetracycline or 25 µg/ml canamycin). Bacteria are inoculated into 2 - 2,5 ml of LB medium with or without antibiotic and placed on night in an incubator at 37°c under shaking. This preparatory culture serves disposing stab or inoculating voluminous (20 – 500 ml) bacterial cultures.

7.1.1.3. Generating competent bacterial cells

1 ml of a preparatory culture will be inoculated into 200ml of LB medium and shacked in the incubator at 37°c to reach an OD_{600} of $0,5 – 0,7$. The culture is then placed on ice for 30 min and then centrifuged for 10 min at 4°c and 3000 rpm. The pellet will be resuspended in 120 ml of a cold CalCl2 solution 0,1M and placed for 30 min on ice. The cells are one more time centrifuged and the pellet is resuspended in 8 ml of CalCl2 solution 0,1M, and can be stored in this form at 4°c for one week or as a glycerol stock (60% glycerol) at -80°c for several months.

7.1.1.4. Bacterial culture for plasmid-DNA preparation

$0,2 – 1$ ml of a preparatory culture is inoculated into $100 – 200$ ml of LB medium with or without antibiotic and grown in the incubator at 37°c to reach an OD_{600} of $0,5 – 0,7$. At this point the culture is treated with chloramphenicole (350 µg/ml) and incubated for 12h in order to stop cell division and enhance plasmid replication [Clevell et al. 1978]. Cells are then centrifuged (20 min, RT, 4000 rpm), and harvested.

7.1.2. Working with RNA

7.1.2.1. Precautions to be taken

Because of its sensibility toward Rnases, working with RNA requires specific carefulness. According to this, all solutions and units used while dealing with RNA must be heated at 240°c for 4h or treated with DEPC (0,1% end concentration) (Sambrook et al. 1989)

7.1.2.2. RNA preparation from cultured cells

HeLa cells were cultivated to confluence in 55 cm^2 culture dishes. DMEM was removed and cells were washed two times with 5 ml of PBS, treated with 600 µl of ß-mercaptoethanol and transferred into eppendorf caps after they have been scraped from the culture dishes. After lysing the cells by passing them several times through a G20 needle using a syringe, 600 µl of ethanol 70% were added to the lysates. The solution was mixed, loaded onto RNAsy column, centrifuged for 15s at 8000*g and the flow-through removed. To wash the RNA, the column was loaded with 700 µl of RW1 buffer and centrifuged as above. The column was transferred onto a new recipient and the RNA washed 2 times with 500 µl of RPE buffer. The first time by centrifuging as for 15s at 8000xg and the second time by centrifuging at 14000xg for 2 min. Finally, the RNA was eluted with 30 µl of RNAse free water.

7.1.2.3 Northernblot analysis

Total RNA (5-15 µg) isolated form normal control and RNAi cells using the RNAesy Kit from QIAGEN (Hilden, Germany) according to the manufacturer instructions was separated in a formaldehyde gel (3%) gel by electrophoresis and transferred to Hybond Nylon membrane (Amersham Pharmacia Biotech). The Northernblot was performed as described in Sambrook et al. (Sambrook et al. 2002). The blots were prehybridized in Church buffer (Sambrook et al. 2002), and hybridized with a ^{32}P-labelled cDNA probe (labeling was performed using Megaprime Labelling Kit, Amersham Bioscience) at 50°C for 16h. The membrane was washed 2x in 2x SSC for 20 min at RT and 0.1x SSC / 0.1% SDS at 68°C and analyzed with a Fuji Image Analyzer BAS1000 (Fuji Photo Film Co.Ltd, Tokyo,Japan).

7.1.2.4 In vitro dsRNA generation.

To generate *schlank* dsRNA, a fragment of the *schlank* cDNA was generated by PCR reaction from LD18904 containing the entire *schlank* cDNA using specific *schlank* primer pairs containing T7 sites and cloned into the Topo pCRII vector (Invitrogen Corp., Carlsbad, CA). The following primers were used:

Schlank RNAiF1:   CTTTTAATACGACTCACTATAGGGAGACGTTTCTGGATA
TCGCCCGTAGGCAAATC
Schlank RNAiR1:   GTTTTAATACGACTCACTATAGGGAGACAGACCCTTCTGCAGAG
AGTCGACAACAA

DsRNA was synthesized using RiboMax Large Scale RNA Production System – T7 – kit (Promega) according to the supplier's instructions. After synthesis, annealing and RNAse (specific for single stranded RNA)/ DNAse treatment, dsRNA was purified by chloroform/ isoamylalcohol extraction and precipitated using standard natrium-acetate/isopropanol precipitation. The dsRNA was analyzed on an 1% TAE-agarose gel to ensure that at least 80-90% of the RNA are double stranded (Schepers, 2004). Micromolar concentration was determined with Oligonucleotide Properties Calculator (Northwestern University).

7.1.3. Working with DNA

7.1.3.1. Phenol extraction and Ethanol precipitation

Proteins are eliminated from DNA probes by phenol extraction (Sambrook et al. 1989). DNA solutions are solvent extracted 2x with 1 volume of phenol eq. and 2x 1vol phenol mix. In order to completely remove phenol traces, the DNA solution is treated 2x with 1vol of chloroform/isoamyl alcohol (1:1). After every extraction step, phase separation is acheaved by centrifugation at RT for 5 min at 13000xg. The purified DNA solution is treated with 0,1vol NaAc 3M (pH 5,3) and precipitated with 2,5 ml vol ethanol 100% (1h, -20°c or 15 min at -80°c). DNA is then pelletized by centrifugation for 10 min at 4°c and 13000xg) and washed with 20-100 µl of ethanol 70% to remove salts.

7.1.3.2. Preparation of plasmid DNA from bacterial cells

The preparation of purified plasmid DNA from 1-200 ml bacterial culture suspension is performed by alkali lysis (Birnboim and Doly, 1979) followed by anion exchange chromatography (plasmid preparation according to QIAGEN). Bacterial cells are centrifuged for 10 min at RT and 4000rpm and the pellet is resuspended in 0,25-4 ml of lysing buffer P1. After breaking the cells by adding to the solution of 0,25-4 ml of buffer P2 and incubating for 5 min at RT, proteins and chromosomal DNA are precipitated by adding of 0,35-4 ml of neutralising buffer P3 and incubation for 10 min at 4°c. Finally, cell detritus are down centrifuged at RT for 20 min at 14000xg and the supernatant is purified by anion exchange chromatography (QIAGEN) or by phenol extraction.

7.1.3.3. RT-PCR using total RNA

Using mRNA as template, the retroviral enzyme reverse transcriptase is able to synthetize a complementary DNA string (cDNA). Total RNA is previously denatured by heating at 70°c for 10 min and then immediately placed on ice to avoid renaturation.

To perform RT-PCR, two master mixes were prepared:

Master mix 1:
          300 nM dNTPs
          20 pmol Primer1
          20 pmol Primer2
          2,5ul DTT (100 mM Stock solution)
          1u RNasin
          1ug total RNA
          total 25 ul RNAse freies Wasser
Master mix 2:
          1x RT-PCR-buffer
          3.5ul Enzym-mix
          Total 25 ul RNAse free Watter

The two master mixes will be unified and RT-PCR performed using the following program:

50°C   30 min Reverse Transcription

94°C   30 s
62°C   45 s   30 cycles
68°C   1min
68°C   10 min

7.1.3.4. PCR (Polymerase chain reaction)

PCR is a procedure for in vitro amplification of DNA by using DNA polymerases from the bacterium *Thermus aquaticus*. To perform PCR, the following master mixes were used:

Mastermix 1:
          300 nM dNTPs
          20 pmol Primer1
          20 pmol Primer2
          Total 25 ul RNAse freies Wasser
Mastermix 2:
          1ul DNA (0,1 – 0,5ug)
          1x buffer 2
          3.5 ul Polymerase
          total 25 ul RNAse free watter

The two master mix will be unified and RT-PCR performed using the following program:

94°C   2 min   Hot start

```
94°C    30 s
62°C    45 s          ⌐
68°C    3min          |    30 cycles
68°C    10 min        ⌐
```

7.1.3.5. DNA dephosphorylation (CIP-Reaction)

The alkaline phosphatase catalyses the elimination of a 5'-phosphate from DNA. After their linearization, plasmids are often dephosphorylated by using the alkaline phosphatase in order to prevent their relegation, while ligating in them a DNA insert. Without the 5'-phosphate unit, the ligation reaction catalyzed by the ligase enzyme cannot happen.

The plasmid DNA is digested for two hours at 37°c with the appropriated restriction enzyme (s. mm). After purification by gel electrophoresis and extraction, it is resuspended in 1x NEB buffer (0,5 μg/ 10 μl) and incubated with the CIP-enzyme (0,5 U/μg DNA) for 1h at 37°c. Finally, the DNA will be purified.

7.1.3.6. Measuring the concentration and the degree of purity of nucleic acids

The concentration and the degree of purity of nucleic acids are obtain by photometric measurement at 260 and 280 nm (Sambrook et al. 1989).

7.1.4. Agarose gel electrophoresis

The agarose gel electrophoresis is a standard procedure for separation, identification and purification of DNA fragments [Nathan and Smith, 1975]. The separation of the negatively charged DNA or RNA fragments is based on their migration to the positively charged pole (Anode) within the electrical field. According to the length of the DNA, the separation is performed in a gel of 0.3-3% under an electrical voltage of 1-5V/migrated cm [Winnacker, 1987]. DNA or DNA/RNA hybrids are visualized by ethydiumbromide staining. Ethydiumbromide is able to insert in-between the helical structure of DNA, whereby a stained complex is formed that can be visualized under UV-light (260-360 nm).

7.1.4.1. Separation of DNA fragments

DNA fragments solved in 1x gel loading buffer are separated in agarose gel with a concentration of 0,7-3% at 10-100mA.

7.1.4.2. Extraction of DNA from agarose gel slice

After separating DNA in a gel electrophoresis and identifying the bands with the fragments of interest, these are cut out from the gel and the slices weighted, and transferred into Eppendorf vessels. To solve the gel slices, QG buffer (300 μl/100 mg of gel slice) is added and incubated for 10 min at 50°c. For DNA fragments with 5kb or more, isopropanol is added after the slice has been solved. The solution is then transferred onto a MinuElut coloumn that is applied onto a 2ml collecting vessel and centrifuged for 1 min at 14000xg. The through flow is discarded and the column loaded with 500 μl of QG buffer and centrifuged as above. The through flow is discarded and column loaded with 720 μl of PE buffer and centrifuged again. The through flow is discarded and column centrifuged again to remove any traces of PE buffer. Finally, the DNA is eluted with 30 μl of Elution Buffer.

7.1.5 Cloning of DNA fragments

7.1.5.1 Enzymatic digestion of DNA

Restriction endonucleases of type II are indispensable in analysing and cloning of DNA fragments, because of their ability to cut DNA at specific recognizing sites, leading to sticky or to blunt ends. The conditions of the enzymatic restriction of DNA vary, according to the used enzyme, whereby the choice of the incubation temperature and reaction buffers is made, referring to the manufacturer indications. For analytical purposes, 0,1-0,5 μg of DNA in a total volume of 20-25 μl with 0,5-1μl of enzyme are incubated for 1-5h at 37°c. Preparative probes with 1-5μg of DNA are digested on night in a total volume of 20-50μl with 5-10 μl of enzyme.

7.1.5.2. DNA ligation

T4-ligase catalyses the formation of a phosphoric di-ester bond between the 5'-phosphate and the 3'-OH of two double stranded DNA fragments. 20-50ng of plasmid DNA and 2-100 times exceeding masse of the insert, are digested with the same restriction enzyme in order to generate compatible ends. To perform a ligation reaction, the following solution mix is prepared and incubated at RT for 40 min.

Reactants	Quantity
Plasmid	10-100 pmol
Insert	10-100 pmol
5x dilution buffer	2μl
2x ligation buffer	10μl
Ligase	8u

7.1.5.3 DNA Transformation

To perform transformation, 1/10 volume of the ligation solution is mixed to 40-200μl of competent cells and incubated for at least 30 min on ice. The cells are then heat shocked at 42°c for 45s-2 min, resuspended into 1 ml of LB-medium and rotated for 1h at 37°c. Aliquots (50-100μl) of this cell suspension are plated on an agar plate (LB-medium, 1,5% agar,100µg/ml of ampicillin, 12,5 µg/ml of tetracycline, or 25µg/ml of tetracycline) and incubated for 12h at 37°c.

7.2 Cell culture

7.2.1 Culture conditions

Mammalian cells. Mammalian cell culture was performed in DMEM (Invitrogen) supplemented with 10 % FCS (Biochrom) and 1% of penicillin/streptomycin at 37°C and 5% CO2. All works related to cell culture are made under absolute sterile conditions.

Drosophila Schneider cells (S2). S2 cells were propagated in Schneider's medium (PAN) supplemented with 10% heat inactivated fetal bovine serum and 1% penicillin/streptomycin (Sigma, St Louis, USA) and cultured at 22°C.

7.2.2 Harvesting of cultured cells

Adherent monolayer Mammalian cells. Culture medium is removed and cells are washed 2x with sterile PBS. Cells are treated with a Trypsin/EDTA solution (0,1%) and incubated at 37°c for 3 min. After cells are detached from the bottom of the culture vessel, trypsinisation is stopped by adding an appropriated volume of DMEM containing 10% of FCS. The cell suspension is centrifuged for 5 min at 2100xg and the pellet is resuspended in an appropriated volume of DMEM containing 10% of FCS, whereby an aliquot is used for further culture.

Drosophila Schneider cells. Cells are detached by simply strong shaking of the culture dishes. The cell suspension is centrifuged for 5 min at 2100xg and the pellet is resuspended in an appropriated volume of Schneider's medium (PAN) containing 10% of FCS, whereby an aliquot is used for further culture.

7.2.3 Cell freezing

Mammalian cells. Harvested mammalian cells lines are resuspended in a solution containing 50% FCS, 40% DMEM and DMSO (10%). Aliquots of 1ml of the cell suspension are transferred into freezing caps and placed in liquid nitrogen.

Drosophila S2 cells. Harvested mammalian cells lines are resuspended in a solution containing 50% FCS, 40% Schneider's medium (PAN) and DMSO (10%). Aliquots of 1ml of the cell suspension are transferred into freezing caps and placed in liquid nitrogen.

7.2.4 Transfecting mammalian cells with DNA

To introduce the expression plasmid into mammalian cells, we used the transfection reagent FuGENE (Roche) according to a modified protocol of the manufacturer. HeLa cells were grown to 50-60%

confluency. The culture medium was removed and cells were washed with 5 ml PBS, detached by trypsinisation and resuspended in 3 ml of serum containing DMEM. To 100 µl of serum free DMEM in a 1,5ml vessel 3 or 6 µl of FuGENE reagent were added, mixed throughly and incubated for 5 mn at RT. 1 or 2µg of DNA were added to solution and incubated again for 15 mn at RT. 1ml of the cell suspension was added to the solution, vigorously schaked and seeded into 6well plates. After incubating the cells for 10 mn at RT, 2ml of DMEM supplemented with 10 % FCS and penicillin/streptomycin (1%) and incubated at 37°C and 5% CO_2 for 24h. Cells were then selected against G418as follows: 1.-5. day 1µg/ml; 6.-10. day 2µg/ml; 11.-15. day 5µg/ml; 16.-20. 30µg/ml

7.2.5 dsRNA Transfection in S2 cells

$2x\ 10^6$ S2 cells were propagated in 1 ml of serum free S2 medium contained in 6well plates and allowed to attach for 30 min at RT. Then *schlank*-dsRNA was added drop wise while slowly rotating the plate, and cells incubated for 1h at 22°C. Thereafter, 2 ml of S2 medium containing inactivated FCS and Pen/Strep (final concentration 10% and 1% respectively) were added and cells incubated at 22°C for 4-5 days. Then, the medium is replaced by fresh medium containing dsRNA and cells further incubated for other 4-5 days to allow a complete lipid turnover.

7.3 Protein chemical methods

7.3.1   Protein concentration measurement (Bradford)

The Bradford's method is a quick procedure for measuring protein concentration (Bradford 1976). The method is based on the shift of Coomacie blue G-250 extinction maxima from 465 to 595nm due to protein binding. 200 µl of Coomacie blue G-250 (1% Coomacie, 8,5% phosphoric acid, 5% ethanol) are added to 20 µl of a protein probe, whose concentration must be measured and the extinction is read at 595 nm against a BSA (0,25-5mg) calibration curve.

7.3.2 SDS-PAGE

Proteins and immune complexes are separated in continuing gels and in denaturing and reducing gels of 12,5-16% , according to Lämmli. An appropriated volume of Lämmli buffer (denaturing buffer) is added to the protein probe and heated at 95°c for 10 min. The electrophoresis is performed at 20mA for 4h in vertical, 6x8 cm large and 0,5 mm thick gels or vertical, 16x14 cm large and 0,7-1 mm thick gels respectively

7.4. Immunocytochemistry

Double immunofluorescence labeling of pSAP and the cytosolic protein ß-adaptin was performed. Normal control and pSap-RNAi HeLa cells grown on glass coverslips were washed with PBS+ (1x PBS, 0.1 M $CaCl_2$, 0.1 M $MgCl_2$), fixed in 4% of paraformaldehyde and permeabilized with 0.1%

Triton X-100 in PBS+. Cells were incubated for 1h at ambient temperature with primary antibody, mouse monoclonal anti-rat ß-adaptin antibody (1:1000 in 0.1% Triton, 1% BSA in PBS+) and washed three times with PBS+, then with the rabbit anti-human Sap-D antibody (1:500 in 0.1% (vol/vol) Triton, 1% BSA in PBS+). The mouse monoclonal anti- rat ß-adaptin antibody was generously provided by T. Kirchhausen (Harvard Medical Scool, Boston, USA); the rabbit polyclonal anti-human Sap-D antibody was made in K. Sandhoff's laboratory. Cytosolic ß-adaptin was visualized using Alexa 594-conjugated goat anti-mouse secondary antibody (Molecular Probes) (1:1000) in 1% (vol/vol) of PBS+ (1x PBS, 0.1 M $CaCl_2$, 0.1 M $MgCl_2$) applying the same conditions as described for the primary antibody. The Sap-D antibody was detected with Alexa 488 conjugated goat anti-rabbit secondary antibody. Finally, cells were washed with PBS and mounted on the microscope slide with antibleach mounting medium (25% phenylendiamin, 80% glycerol)

7.4 Microscopy and Image Analysis

Fluorescent microscopy was performed with a confocal laser scanning unit (LSM) coupled to a Zeiss Axiovert S100 (Laser lines: Argon 488 nm and HeNe 546nm; filters for green fluorescence bandspan 505-550 nm, for red fluorescence langspan 560 nm, line averaging 2x, 4x). Images were recorded with equal exposure times for specific antibodies. In each experiment, at least 10 coverslips were analyzed and gene expression was quantified in both control and targeted cells by visual counting.

7.5 Sphingolipid analysis in mammalian cells

7.5.1 Pulse chase labeling with ^{14}C-Serine

Metabolic labelling and lipid extraction were performed as described previously (Klein et al. 1994). Wild type and RNAi cells were cultured to confluence in $25m^2$ cell culture flasks. Medium was removed and cells were washed twice with 2,5 ml of serum free MEM containing 1% penicillin/streptomycin. 2,5 ml of MEM supplemented with 1% pen / strep, 0.3% FCS and ^{14}C-serine (1µCi/ml) were added and cells were incubated for 24h at 37°C /5% CO2.

Chase. Pulse medium was removed and cells were washed 2x with 2,5 ml of serum free MEM containing 1% pen/strep. 5ml of chase medium (MEM supplemented with 1% pen / strep, 0,6% FCS and 10x excess of unlabeled serine) were added and cells were incubated for 120h at 37°C and 5% CO2. After 120h chase, medium was removed and cells were harvested and lipids extracted as indicated above.

7.5.2 Lipid extraction

RNAi and normal control cells grown to confluency were washed twice with PBS and harvested by trypsinization. After centrifugation for 10min at 2000rpm, supernatant was removed and cells were washed twice with PBS and frozen at –20°C for at least 2h. Cells were resuspended in 1ml of water, and lysed by sonicating 4x 30s with 10s of interval. 2 ml of methanol was added to the lysates and sonicated in water bath for 15min. After the addition of 2ml of methanol and 3ml of chloroform/methanol/water, 50/10/1, v/v/v, lipids were extracted for 12hrs at 50°C. To remove cell debris, samples were centrifuged at 4000 rpm for 15 min and supernatants were transferred into new glass tubes. Remaining cell debris were resuspended in 2 ml of C/M; 1/1; v/v, sonicated for 5mn, centrifugated and the supernatant added to the first ones. Samples were then dried in a nitrogen stream.

7.5.3 Alkaline methanolysis

2.5 ml of methanol were added to the dried lipids and after sonication for 5min, 25µl of 5M NaOH were added. After incubation for at least 2h at 37°C, samples were neutralized with 7µl of acetic acid, and dried in a nitrogen stream.

7.5.4 Reversed phase chromatography

Column preparation. 2ml of RP-18 (40-63µm) in methanol were loaded in Pasteur pipettes blocked with glass wool, and equilibrated with:

Water	2ml
Methanol	1ml
C/M 1/1	2ml
Methanol	1ml
C/M/W 0,1MKCl, 3/48/47	2ml

Dried lipid samples were dissolved in 1ml of methanol and sonicated for 5 min. Then 1ml of ammonium acetate (300 mmol) was added and samples were loaded onto RP18 (40-63µm) columns (2ml). Residual Lipid in the pyrex glass tube was recuperated with 2ml of methanol/200mM ammonium acetate (1/1,v/v) and as well loaded onto the RP18 columns. After desalting the samples with 6x 1ml of water and equilibrating the columns with 400 µl of methanol, glycosphingolipids were eluted with 8x1ml of chloroform/methanol, 1/1, v/v and dried in a nitrogen stream.

*Separation of neutral and acidic glycosphingolipids.*To elute the neutral glycosphingolipids, dried samples were dissolved in 1ml of chloroform/methanol/water (3/7/1, v/v/v) and applied onto columns containing 1,5ml of equilibrated DEAE-sepharose and the flow through recuperated. Remaining lipids in the glass tubes were recuperated by washing with 2 ml of chloroform/methanol/water (3/7/1, v/v/v) and as well loaded onto columns. The remaining neutral glycosphingolipids were then eluted with 4ml of chloroform/methanol/water (3/7/1, v/v/v) and dried in nitrogen stream.

Gangliosides were eluted with 8ml of chloroform / methanol / 1M ammonium acetate (3/7/1, v/v/v), dried under nitrogen stream and desalted as described above.

TLC. Aliquots corresponding to equal amounts of total cell proteins were separated on TLC plates (HPTLC, Merck, Darmstadt) using C/M/ CaCl$_2$ (15 μmol), 60/35/8, v/v/v or C/M/ Water, 65/25/4 v/v/v for neutral glycosphingolipids and C/M/ CaCl$_2$ (15 μmol), 55/45/10, v/v/v for acidic glycosphingolipids and C/M/Acetic acid, 190/9/1, v/v/v for ceramide. Sphingolipids spots were visualized by exposing to a Phosphoimager and analyzed with a Fuji Image Analyzer BAS1000 (Fuji Photo Film Co.Ltd, Tokyo, Japan) (Schepers et al. 1996).

7.5.5 Feeding cells with recombinant prosaposin

Culture medium was removed from confluent wild type and pSap-RNAi HeLa cells and replaced with fresch medium culture medium containing containing 0,3-0,2 μg/ml of recombinant prosaposin and incubated for 24h. Cells were then harvested, their sphingolipids extracted and analyzed by TLC.

7.5.6 Densitometric quantification of glycosphingolipids

Sphingolipids were extracted from unlabelled wild type and RNAi cells as described above and aliquotes corresponding to equal amounts of total cell proteins were separated on HPTLC plates. To detect neutral and acidic glycosphingolipids, plates were sprayed with orcinol (Bial's reagent) and resorcinol respectively, and heated at 100°C for 10 min. Ceramide was visualized by spraying the plate with CuSO$_4$ in H$_3$PO$_4$ and heating at 180°C for 10 min.

7.5.7 Metabolic labeling with [14C]-Galactose

Normal control and GCS-RNAi HeLa were grown to 80 - 85% confluence in 9 mm culture dishes. Culture medium were removed and replaced by a new culture medium containing D-[^{14}C]-Galactose (2μCi/ml, specific activity 57, mCi / mmol) and incubated for 10h before harvesting. The labeled sphingolipid fraction was isolated and, separated by HPTLC and visualized by phosphoimaging.

7.5.8 Metabolic labeling with [3H]-sulfatide

WT and pSap-RNAi HeLa cells were incubated with [^{3}H]-sulfatide (0,25 nmol /ml) for 48h. Subsequently, the medium was changed and the cells were incubated with MEM containing 10% FCS and 1% of penicillin/streptomycin for additional 48h. After harvesting the cells, the glycosphingolipid fraction was extracted and aliquots corresponding to equal amount of total cell proteins were separated on TLC. Radioactive spots were detected by exposing to a Phosphoimager screen and analyzed with a Fuji Image Analyzer BAS1000 (Fuji Photo Film Co.Ltd, Tokyo, Japan) (Schepers et al. 1996).

7.5.9 Metabolic labeling with NBD-C6 -GM3

Confluent cells in 8 cm^2 cell culture dishes were incubated at 37°C with 10^{-5}M of NBD-C6 -GM3-BSA complex for 6h. After removing the medium, cells were washed 3x with PBS and harvested by trypsin/ EDTA (0,1%). NBD-C6 -GM3 was extracted, separated by TLC, visualized under UV-light and subjected to densitometric quantitation.

7.6 Lipid analysis in Schneider's Drosophila S2 cells and larvae

7.6.1 Metabolic labelling of S2 cells

After treatment with dsRNA for up to 11 days the medium was replaced by fresh serum-free medium containing dsRNA (75 nM) and [^{14}C]serine (1 μCi/ml) or [^{14}C]-Acetat (1,5 μCi/ml) for 24h. The biosynthetic labelling, the extraction of the lipids as well as TLC analysis were performed as described previously (Diallo et al., 2003). For the separation of ceramides and dihydroceramides, TLC plates impregnated with borate were used and developed in chloroform/methanol 90:10 (v/v). Radioactive spots were evaluated and quantified using the bioimaging analyzer Fuji-Bas 100 and the TINA 2.08 software (Raytest, Straubenhardt, Germany).

7.6.2 LC/APCI-MS analysis

Lipid extracts of S2 cells were analysed by means of LC/APCI-MS as previously described (Farwanah et al 2003). However, the LC system was a 2695 Alliance separation module (Waters, Eschborn, Germany). In addition, The LC system was coupled to a Q-TOF 2 (Waters, Eschborn, Germany) mass spectrometer. The mass spectra were performed in the full scan mode (m/z: 400-1500). The collision energy was 50V.

7.6.3 Dihydroceramide synthase assay.

The S2 were lysed in buffer A (50 mM Hepes/NaOH, pH 7.5, 0.5 mM DTT, 2x protease inhibitor (Complete^{TM}; Roche, Indianapolis, USA)) by sonication, and centrifuged at 10,000 rpm for 10 min in a tabletop centrifuge at 4 °C. The supernatant was transferred to a new tube. Protein content was measured with 4 μl and 2 μl of the supernatant combined with 46 μl and 48 μl buffer A in 950 μl Pierce Assay reagent (BCA Protein Assay Kit, Pierce). Following incubation for 30 min at 37 °C, OD$_{260}$ was determined and compared with a standardization curve. 40 μg of protein were mixed with 5 μM dihydrosphingosine (Sigma, Germany), 0.2 μCi of [4,5-^{3}H] D-erythro-dihydrosphingosine (82 Ci/mol) and 25 mM fatty arachidoyl CoA in buffer B (50 mM Hepes/NaOH, pH 7.5, 0.5 mM DTT, 1 mM MgCl$_2$, 0.1 % digitonin) in a volume of 100 μl. After incubation for 20 min at 37 °C, the reaction was stopped with 100 μl of chloroform/methanol (1:1, v/v). The dried lipids were resuspended in 40 μl of chloroform/methanol (1:1, v/v) and separated on Silica Gel60 HPTLC plates (Merck). The separated lipid bands were visualized with a phosphoimager (FUJIX Bio Imaging Analyser 1000, Fuji Photo Film, Tokyo, Japan) and quantified by TINA 2.08 software (Raytest, Straubenhardt, Germany).

7.7 Materials

7.7.1 Equipments

Autoclav   Fedegari, Pavia, Italien
Incubators   Heraeus-Christ Hanau
Cover slips   Menzel Gläser, Breda, Niederlande
ELISA-Photometer   Titertek Multiscan Plus MKII, Fa Flow
Gel equipments   Horizontalelektrophoresessystem Horizon, Gibco-BRL
        Protein gel elektrophoresis, Biorad, München
        Hoefer Scientific Instruments, San Francisco, USA
        Blotting equipments Firma Biorad, München
Heatin block   Dri-Block, DB3Techne
Magnetic tirrer   Ika-Combimag RCT, Ika Werk, Staufen
Membranes   Boehringer, Mannheim
        QIAGEN, Hilden
        Biorad
Microskop   Wilovert S, Hund, Wetzlar
        Fluorescent microskop Zeiss Axiovert 35 mit Filtern (Bp 485, Ft 510, 515-565),
        Oberkochen
Microwave   Microchef FM 3915, Moulinex
Object holder   Superfrost Plus, Menzel Gläser, Breda, Niederlande
PCR-Thermoblock   MJResearch, Newton Massachusetts
pH-Meter   Typ Digital350, Beckmann, Großbritannien
Phosphoimager   Fuji Bas 100 und Auswertungsprogramm TINA Version 2.08, Raytest,
        Straubenhardt
Photometer   Pharmacia Biotech, Uppsala, Schweden
Pipettes   Gilson Pipetman P20, P200, P1000
        Abimed, Langenfeld
        10µl, Eppendorf Reference, Hamburg
X-Ray protective clothing   Chronex, Du Pont de Nemour, Frankreich
X-Ray radiographs   X-Omat AR5, Biomax TM, Eastman Kodak, NY, USA
Rotors   Typ GSA-Sorvall, Du Pont, Frankreich
        Typ SW28, SW41, Beckmann, München
Shaking incubator   New Brunswick Scientific, New Brunswick, Großbrittannien
Shaking watter bath   Gesellschaft für Labortechnik, Burgwedel
Voltage equipment   Power Supply Typ 2197, 2103, Macrodrive 5, LKB-Pharmacia, Uppsala,
Schweden
Clean bench   Biohazard, Gelaire, Mailand, Italien
        Heraeus-Christ, Hanau
Oven   Memmert, Schwabach
Scintillation counter   TRI-CARB 1900 TR, Canberra Packard
Ultrasonic device   Sonifer B12, Branson Sonic Power company, USA
UV-Transilluminator   Biometra, Göttingen
Vortex devicet   Bender und Hohbein, Zürich, Schweiz
Watter cleaning   Millipore-Pelicon Filtrationsanlage mit Polysulfonfilterkassette (PTGC, 10000
        MW), Millipore, Molsheim, Frankreich
Centrifuges   RC5-B Sorvall-Kühlzentrifuge, Sorvall
        Ultrazentrifuge L8-80, Beckmann, München
        Eppendorf-Tischzentrifugen 5414, Varipette 4710, Eppendorf, Hamburg
        Minifuge GL, Heraeus-Christ, Hanau
        Megafuge 2.0 Heraeus-Christ, Hanau

7.7.2 Steril Plastic Materials

Tissue culture flasks Falcon, GB
Tissue culture vials   Costar, USA
                        Greiner, Nürtingen
Petri dishes   • 90 mm, • 135 mm, Falcon, Großbritannien
Polyallomer tubes   Beckmann, München
Reaction vessels   1,5 ml, 2 ml, Eppendorf, Hamburg
                        0,5 ml ultradünn, Biozym, Oldenburg
                        12,5 ml PPN, 15 ml, Greiner, Nürtingen
                        50 ml Falcon / Costar, Großbritannien
Pipett accessories   Abimed, Langenfeld
                        Greiner, Nürtingen
Cell culture flasks   25er,75er,162er, Costar USA
Aseptic filters FP 030/3, Schleicher und Schüll

7.7.3 Antibodies

anti-Rabit- IgG –Antibodies
HRP-conjugated   from New England Biolabs (USA)
Got-anti-Human Sap freundlicherweise zur Verfügung gestellt von Pr. R. Pagano, USA

7.7.4 Enzymes

Restriction endonucleases   New England Biolabs, USA
Reverse Transkriptase   Roche Mannheim
(Titan one tube)

    RNAse Inhibitor   Promega, Heidelberg
    T7, RNA-Polymerase

    (T7 Ribomax)   Promega, Heidelberg
    Taq-Polymerase   Expand-Long-Template, Boeringer, Manheim
    Calf intestinal

    phosphatase (CIP)   New England Biolabs, USA

    Trypsin   Roche, Manheim

    T4 DNA Ligase

    (Rapid Ligation Kit)   Roche, Mannheim

7.7.5 Radioaktivity

$[^{14}C]$-Serin   Amersham
$[^{3}H]$-sulfatide   from research group
$[^{14}C]$-Galactose from Amersham Pharmacia Biotech. (Rainham, Essex, UK)
$[4,5-\,^{3}H]$ D-erythro-dihydrosphingosine

7.7.6 Bakterial strains

Escherichia coli: DH5α, DH10Bac

7.7.7 Cell lines

Labor appellation	Cell types	Provenance
HeLa-cells	ovarian tumor	from research group
Human fibroblasts	Wild type	from research group
Human fibroblasts	pSap-/-	from research group
Mouse fibroblasts	Wild type	from research group
Mouse fibroblasts	pSap-/-	from research group
Mouse melanocytes	MEB4	from Van Meer's research group
D.m embryonic cells	S2	from research group

7.7.8 Media

LB-Medium for Bacteria
DMEM +10% FCS + 1u/ml Penicillin/Streptomycin for HeLa Cells
DMEM + 1u/ml Penicillin/Streptomycin für HeLa Cells
DES Expression Medium for S2 Cells
DES serum free Medium for S2 Cells

7.7.9 Plasmids

pBluescript SKII(-)	Stratagene, Heidelberg
pcDNA3.1	Invitrogen

7.7.10 Manufacturers

All restriction senzymes, Rainbow standard protein 756, IPTG, X-Gal, Ampicillin, DNAse, Proteinase K (*New England Biolabs, USA*)
Penicillin/Streptomycin, Tiamolin, Minocyclin (*Biochrom*)
Ampicillin, dNTP-Mix, Kanamycin, Expand-Long Template Taq-Polymerase; Rapid Ligation Kit Acrylamid solution. 30%, Nylon membranes, T7-Polymerase, Titan-one-tube (*Roche, Mannheim*)
Agarose, Triton x-100, ; 1KB DNA standard ladder, FCS, MEM-Medium, G418, pcDNA3.1 (*Gibco BRL, Bethesda*)
MEM-Medium,
PVDF-Membran (Macherey und Nagel, Düren)
Ammonium acetat, Bromophenol blue, Dinatriumhydrogenphosphat, EDTA-Na-Salt, Glycerol, Isoamyl alcohol, Potassium acetat, Potassium chlorid, Potassium dihydrogen phosphat, N,N´-Dimethyl bisacryl amid, Sodium acetat, Sodium chlorid, Sodium hydroxid, , RNAse-Inhibitor, T7 Ribomax Kit (*Promega*)
DNA purifying mini columns: QIAPrep Midi-, Mini-, Spin-Kit, QIAex-Agarose elution Kit, RNeasy-RNA preparation kit, Nylon membranes, (*QIAGEN, Hilden*)
All Salts and organic solvents, Formaldehyd, Formamid, Hydrochloric acid, Acetic Acetic, (*Riedel de Haen, Seelze*)
Blotting paper, Whatmann 3M, Filterpapiere (*Schleicher und Schüll, Dassel*)
Acrylamid, Ethidium bromide, PVDF-Membran Coomassie R250 (*Serva, Heidelberg*)
Acrylamid, N,N´-Bisacrylamid, Ethidium bromide, HEPES, Sodium dodecyl sulfate, Kodak- Biomax X, Kodak GBX developer and X-Ray fixer, TEMED, ☐-Mercaptoethanol, Tween 20, Triton X100 Chloramphenicol Synthetic Oligonucleotides (*Sigma, Taufkichen*)
Plasmids pBluescript KS (+) (*Stratagene*)

7.8 Buffers and Solutions

AA-Stock solution.3 (38.5%)

37.5	%	Acryamid
1	%	Bisacrylamid

Buffer A
50 mM Hepes/NaOH, pH 7.5
0.5 mM DTT
2x protease inhibito

Buffer B
50 mM Hepes/NaOH, pH 7.5
0.5 mM DTT
1 mM MgCl$_2$
0.1 % digitonin

Standard BSA (Bradford)

0.025	%	BSA
0.1	N	Formic acid

Chloroform/ Isoamylakohol

96	%	CHCl$_3$
4	%	Isoamylalkohol

Coomassie Stock solution (5x)

0.33	%	ServaBlue G
17	%	EtOH
27	%	H$_3$PO$_4$

Strip solution

25	%	Methanol
10	%	Acetic acid

Staining solution (Coomassie)

0.1	%	Coomassie Brilliant Blue
		R250
25	%	Methanol
10	%	Essigsäure

Fixiation solution

25	%	2-Propanol
5	%	Acid acetic
1	%	Glycerol

Gel loading buffer II (10x)

25	%	Ficoll
1	mM	EDTA (pH 8.0)
0.25	%	Bromphenolblau
0.25	%	Xylencyanol

Gel loading buffer III (2x)

95	%	Formamid
20	mM	EDTA (pH 8.0)
0.05	%	Bromophenol blue

Gel loading buffer IV (5x)

312.5	mM	Tris-HCl (pH 6.8)
2.5	%	SDS
5	%	□-Mercaptoethanol
50	%	Glycerol
0.025	%	ServaBlue G

IPTG 200   mM   in Wasser

LB-Medium

1	%	NaCl
1	%	Trypton
0.5	%	Hefe-Extrakt

SDS-PAGE running buffer

0.1	M	Tris-HCl (pH 8.25)
0.1	M	Tricin
0.1	%	SDS

Lysis buffer I

100	mM	NaCl
25	mM	EDTA (pH 8.0)
0.5	%	SDS
0.1	g/l	Proteinase K (befor use to be added)

Lysis buffer II

50	mM	Tris-HCl (pH7.5)
5	mM	EDTA (pH 8.0)
1	%	Nonidet-P40
0.5	%	BSA
0.05	%	NaN$_3$

Lysis buffer P1

see QIAprep Manual

PAG2

5	%	Acrylamid/ Bisacrylamid
		(AA-Stammlsg.2)
1	x	TBE
0.03	%	TEMED
0.08	%	APS

PBS (pH7.5)

140	mM	NaCl
3	mM	KCl
16	mM	Na$_2$HPO$_4$
1.5	mM	KH$_2$PO$_4$

PCR-Buffer P1 (10x)

17.5	mM	MgCl$_2$
500	mM	Tris-HCl (pH 9.2)
160	mM	(NH$_4$)$_2$SO$_4$

PCR-Buffer P2 (10x)

22.5	mM	MgCl$_2$
500	mM	Tris-HCl (pH 9.2)
160	mM	(NH$_4$)$_2$SO$_4$

PCR-Buffer P3 (10x)

22.5	mM	MgCl$_2$
500	mM	Tris-HCl (pH 9.2)
160	mM	(NH$_4$)$_2$SO$_4$
20	%	DMSO (v/v)
1	%	Tween' 20 (v/v)

Stacking gel (4%)

4	%	AA-Stammlsg 3
1	M	Tris-HCl (pH 8.45)
0.1	%	SDS
1	µl/ml	TEMED
0.1	%	APS

SOB-Medium

2	%	Trypton
0.5	%	Yeast-Extract
0.05	%	NaCl
10	mM	MgCl$_2$
10	mM	MgSO$_4$

SOC-Medium

1	x	SOB-Medium
20	mM	Glucose

TAE (50x)

2	M	Tris•Acetat
50	mM	EDTA (pH 8.0)

TE (pH8.0)

10mM	Tris-HCl (pH7.4)	
1	mM	EDTA (pH 8.0)

Transfer-Buffer (pH 11.0)

10mM CAPS pH 11
10% MeOH

Separating gel (12.5-16.5%)

12-.5-16.5	%	AA-Stammlsg 3
1	M	Tris-HCl (pH 8.45)
0.1	%	SDS
1	µl/ml	TEMED
0.1	%	APS

Trypsin/EDTA (pH7.5)

0.25	%	Trypsin
2	%	EDTA
1	x	PBS

X- Gal

100	mM	in DMF

8. Annex

8.1 Sequences

8.1.1 Human GCS

```
   LOCUS       HUMCGA        1637 bp    mRNA           PRI        10-FEB-1999
DEFINITION   Homo sapiens mRNA for ceramide glucosyltransferase, complete cds.
ACCESSION   D50840
VERSION     D50840.1  GI:1350551
KEYWORDS    ceramide glucosyltransferase; glucosyl ceramide synthase.
SOURCE      Homo sapiens cell_line:melanoma SK-Mel-28 cDNA to mRNA.
  ORGANISM  Homo sapiens
            Eukaryota; Metazoa; Chordata; Craniata; Vertebrata; Euteleostomi;
            Mammalia; Eutheria; Primates; Catarrhini; Hominidae; Homo.
REFERENCE   1  (bases 1 to 1637)
  AUTHORS   Ichikawa,S.
  TITLE     Direct Submission
  JOURNAL   Submitted (01-JUN-1995) Shinichi Ichikawa, The Institute of
            Physical and Chemical Research(RIKEN), Glyco-Cell Biology; 2-1
            Hirosawa, Wako-shi, Saitama 351-01, Japan
            (Tel:+81-48-462-1111(ex.6237), Fax:+81-48-462-4690)
REFERENCE   2  (bases 1 to 1637)
  AUTHORS   Ichikawa,S., Sakiyama,H., Suzuki,G., Hidari,K.I. and Hirabayashi,Y.
  TITLE     Expression cloning of a cDNA for human ceramide glucosyltransferase
            that catalyzes the first glycosylation step of glycosphingolipid
            synthesis
  JOURNAL   Proc. Natl. Acad. Sci. U.S.A. 93 (10), 4638-4643 (1996)
  MEDLINE   96209784
  REMARK    Erratum:[[published erratum appears in Proc Natl Acad Sci U S A
            1996 Oct 29;93(22):12654]]
COMMENT     On Jun 2, 1996 this sequence version replaced gi:1325916.
FEATURES             Location/Qualifiers
     source          1..1637
                     /organism="Homo sapiens"
                     /db_xref="taxon:9606"
                     /cell_line="melanoma SK-Mel-28"
     CDS             291..1475
                     /EC_number="2.4.1.80"
                     /note="glucosyl ceramide synthase"
                     /codon_start=1
                     /product="ceramide glucosyltransferase"
                     /protein_id="BAA09451.1"
                     /db_xref="GI:1325917"
                     /translation="MALLDLALEGMAVFGFVLFLVLWLMHFMAIIYTRLHLNKKATDK
                     QPYSKLPGVSLLKPLKGVDPNLINNLETFFELDYPKYEVLLCVQDHDDPAIDVCKKLL
                     GKYPNVDARLFIGGKKVGINPKINNLMPGYEVAKYDLIWICDSGIRVIPDTLTDMVNQ
                     MTEKVGLVHGLPYVADRQGFAATLEQVYFGTSHPRYYISANVTGFKCVTGMSCLMRKD
                     VLDQAGGLIAFAQYIAEDYFMAKAIADRGWRFAMSTQVAMQNSGSYSISQFQSRMIRW
                     TKLRINMLPATIICEPISECFVASLIIGWAAHHVFRWDIMVFFMCHCLAWFIFDYIQL
                     RGVQGGTLCFSKLDYAVAWFIRESMTIYIFLSALWDPTISWRTGRYRLRCGGTAEEIL
                     DV"
BASE COUNT      416 a      375 c      394 g      452 t
ORIGIN
        1 gaggcgaacc ggagcgcggg gccgcggtcg ccccgaccag agccgggaga ccgcagcacc
       61 cgcagccgcc cgcgagcgcg ccgaagacag cgcgcaggcg agagcgcgcg ggcgggggcg
      121 cgcaggccgt gcccgccct tccgtcccca cccccctccg cccttcctc tccccacctt
      181 cctctcgcct cccgcgcccc cgcaccgggc gcccaccctg tcctcctcct gcgggagcgt
      241 tgtccgtgtt ggcggccgca gcgggccggg ccggtccggc gggccggggg atggcgctgc
      301 tggacctggc cttggaggga atggccgtct tcggttcgt cctcttcttg gtgctgtggc
      361 tgatgcattt catggctatc atctacaccc gattacacct caacaagaag gcaactgaca
      421 aacagcctta tagcaagctc ccaggtgtct ctcttctgaa accactgaaa ggggtagatc
      481 ctaacttaat caacaacctg gaaacattct ttgaattgga ttatcccaaa tatgaagtgc
      541 tccttgtgt acaagatcat gatgatccag ccattgatgt atgtaagaag cttcttggaa
      601 aatatccaaa tgttgatgct agattgttta taggtggtaa aaaagttggc attaatccta
      661 aaattaataa tttaatgcca ggatatgaag ttgcaaagta tgatcttata tggatttgtg
      721 atagtggaat aagagtaatt ccagatacgc ttactgacat ggtgaatcaa atgacagaaa
```

```
 781 aagtaggctt ggttcacggg ctgccttacg tagcagacag acagggcttt gctgccacct
 841 tagagcaggt atattttgga acttcacatc caagatacta tatctctgcc aatgtaactg
 901 gtttcaaatg tgtgacagga atgtcttgtt taatgagaaa agatgtgttg gatcaagcag
 961 gaggacttat agcttttgct cagtacattg ccgaagatta ctttatggcc aaagcgatag
1021 ctgaccgagg ttggaggttt gcaatgtcca ctcaagttgc aatgcaaaac tctggctcat
1081 attcaatttc tcagtttcaa tccagaatga tcaggtggac caaactacga attaacatgc
1141 ttcctgctac aataatttgt gagccaattt cagaatgctt tgttgccagt ttaattattg
1201 gatgggcagc ccaccatgtg ttcagatggg atattatggt atttttcatg tgtcattgcc
1261 tggcatggtt tatatttgac tacattcaac tcagggagtgt ccagggtgc acactgtgtt
1321 tttcaaaact tgattatgca gtcgcctggt tcatccgcga atccatgaca atatacattt
1381 ttttgtctgc attatgggac ccaactataa gctggagaac tggtcgctac agattacgct
1441 gtggggtac agcagaggaa atcctagatg tataactaca gctttgtgac tgtatataaa
1501 ggaaaaaaga gaagtattat aaattatgtt tatataaatg ctttaaaaa tctaccttct
1561 gtagtttat cacatgtatg ttttggtatc tgttctttaa tttattttg catggcactt
1621 gcatctgtga aaaaaa
//
```

8.1.2 Mouse Ugcg

```
LOCUS       NM_011673               3719 bp    mRNA    linear   ROD 21-MAR-2010
DEFINITION  Mus musculus UDP-glucose ceramide glucosyltransferase (Ugcg), mRNA.
ACCESSION   NM_011673
VERSION     NM_011673.3  GI:133892814
KEYWORDS    .
SOURCE      Mus musculus (house mouse)
  ORGANISM  Mus musculus
            Eukaryota; Metazoa; Chordata; Craniata; Vertebrata; Euteleostomi;
            Mammalia; Eutheria; Euarchontoglires; Glires; Rodentia;
            Sciurognathi; Muroidea; Muridae; Murinae; Mus.

REFERENCE   9 (bases 1 to 3719)
  AUTHORS   Ichikawa,S., Ozawa,K. and Hirabayashi,Y.
  TITLE     Molecular cloning and expression of mouse ceramide
            glucosyltransferase
  JOURNAL   Biochem. Mol. Biol. Int. 44 (6), 1193-1202 (1998)
  PUBMED    9623774

FEATURES             Location/Qualifiers
     source          1..3719
                     /organism="Mus musculus"
                     /mol_type="mRNA"
                     /strain="C57BL/6"
                     /db_xref="taxon: 10090"
                     /chromosome="4"
                     /map="4 32.0 cM"
     gene            1..3719
                     /gene="Ugcg"
                     /gene_synonym="AU043821; C80537; Epcs21; GlcT-1;
                     Ugcgl"
                     /note="UDP-glucose ceramide glucosyltransferase"
                     /db_xref="GeneID:22234"
                     /db_xref="MGI:1332243"

     CDS             95..1279
                     /gene="Ugcg"
                     /gene_synonym="AU043821; C80537; Epcs21; GlcT-1; Ugcgl"
                     /EC_number="2.4.1.80"
                     /note="ectoplacental cone, invasive trophoblast giant
                     cells, extraembryonic ectoderm and chorion sequence 21;
                     UDP-glucose ceramide glucosyltransferase-like; GCS;
                     glucosylceramide synthase; UDP-glucose:N-acylsphingosine
                     D-glucosyltransferase"
                     /codon_start=1
                     /product="UDP-glucose ceramide glucosyltransferase"
                     /protein_id="NP_035803.1"
                     /db_xref="GI:7106443"
                     /db_xref="CCDS:CCDS18219.1"
```

```
          /db_xref="GeneID:22234"
          /db_xref="MGI:1332243"
          /translation="MALLDLAQEGMALFGFVLFVVLWLMHFMSIIYTRLHLNKKATDK
          QPYSKLPGVSLLKPLKGVDPNLINNLETFFELDYPKYEVLLCVQDHDDPAIDVCKKLL
          GKYPNVDARLFIGGKKVGINPKINNLMPAYEVAKYDLIWICDSGIRVIPDTLTDMVNQ
          MTEKVGLVHGLPYVADRQGFAATLEQVYFGTSHPRSYISANVTGFKCVTGMSCLMRKD
          VLDQAGGLIAFAQYIAEDYFMAKAIADRGWRFSMSTQVAMQNSGSYSISQFQSRMIRW
          TKLRINMLPATIICEPISECFVASLIIGWAAHHVFRWDIMVFFMCHCLAWFIFDYIQL
          RGVQGGTLCFSKLDYAVAWFIRESMTIYIFLSALWDPTISWRTGRYRLRCGGTAEEIL
          /gene="Ugcg"
          /gene_synonym="AU043821; C80537; Epcs21; GlcT-1; Ugcgl"
ORIGIN
        1 cgcgccccgc accccggcggc ccccgcgtcc tcctcccgcg gcagcgctgt ccgcggcggc
       61 cggagcgggc cgggccgggc cagcgggccg ggggatggcg ctgctggacc tggcccagga
      121 gggaatggcc ttgttcggct tcgtgctctt cgtggtgctg tggctgatgc atttcatgtc
      181 catcatctac acccggttac acctcaacaa gaaggcaaca gacaaacagc cgtatagcaa
      241 gctccctggt gtctctcttc tgaagccact gaaggggtg gatcctaacc taatcaacaa
      301 cttggagaca ttctttgaac tggattatcc caaatatgaa gtactccttt gtgtacaaga
      361 tcatgatgat ccagccattg atgtatgtaa gaaattgctt ggaaaatacc caaatgtcga
      421 tgctagatta tttataggtg gcaaaaaggt tggcattaac cctaaaatta ataatttgat
      481 gccagcatat gaagttgcaa aatatgatct catatggatt tgtgatagtg gaataagagt
      541 catcccagac acattaactg acatggtgaa tcagatgaca gagaaagtgg ggttggtcca
      601 cgggctgccg tatgtagccg acagacaagg ctttgctgcc accttagagc aggtatattt
      661 tggaacttca cacccaagat cctatatctc tgccaatgta actggcttca aatgtgtgac
      721 ggggatgtct tgtttgatga ggaaggatgt gctagatcag gcaggagggc tcatagcctt
      781 tgctcagtac attgctgaag attactttat ggccaaagca atagcccgacc gaggttggag
      841 gtttttcaatg tctactcaag ttgccatgca aaactctggt tcgtactcaa tttctcagtt
      901 tcaatccaga atgatcaggt ggaccaaatt gagaattaac atgcttcctg ctacaataat
      961 ttgtgagcca atttcagaat gctttgttgc cagtttaatt attgggtggg cagcccacca
     1021 tgtattcaga tgggatatca tggtcttctt catgtgccac tgcctggcat ggtttatatt
     1081 tgactacatt caactcaggg gtgtccaggg tggcacactg tgtttttcaa aacttgatta
     1141 tgctgtggcc tggttcatcc gtgaatccat gacaatctac attttccctgt cggcattatg
     1201 ggacccgact ataagctgga gaactggtcg ctacaggtta cgctgtgggg ggacagcaga
     1261 ggagatcctg gatgtgtaag aagacctctg tgactgatgg cgcacagtgt ggaatggaag
     1321 tgttataaat tatgtttata gagacacttt ccagggtctc ccttcagtag tttatcacat
     1381 gtatgttttg gtatctgctc tttaatttat ttgcatggca cttgcatctg tgaaaagcaa
     1441 aaaaacaaaa acaaaacatg tctgtagtct tggccaaata gtatatctg tggaagtaga
     1501 gaatcaaggc aattggttct agaggatggt agatttgtttt ttttgttgtt gtttatttt
     1561 ttaaacaaat aaatatatat atatatatat atatatacac acacacacac atatacagga
     1621 ctgctctgca gcagacacta tgacagtatc cttcagtagg gaaagtttct tacctgagtt
     1681 cctctgaatc tgtacgggca gaacagcagc tccatgagag aagtcatggc ctctcccctg
     1741 tgtgaagtaa atgctataca gagagacgcc cgctggctga gaaacttttt catgcttttat
     1801 gcctacctct tgtagttgtt gcagagcaaa tagaagttag aagttgtaat acggtagcta
     1861 ggccttgcaa aaacaaatgg aaaacttttt aaaataataa taataaaaga gctggagtct
     1921 agtatttata taaatctgtg aggtttttta ttttgttttg ggtttttttt tttgtttgt
     1981 ttttttgttt gtttgggttt ttttgttttg ttttttgttt ttggtctcat tacaattgaa
     2041 ccaaaagtgg ttgagattgg tttttaatgg taaccatgaa ttgaaggcat cttttggacc
     2101 aactgtcgtt ggttctgtct tgaaccatat taatcactgt gctgtaatta ggagagctga
     2161 atctttccct ccctgctccc ggctcacccg gcctgcccca tctttccttc gcttttttcca
     2221 gggggtgatg ttgtacaagt ggatgtttttg tagtgtaagg aaaatgcatt tcagacagtc
     2281 acacaggagc tattttctta cccgggatgt ctgtattgtt aataagaatg taatttcatg
     2341 atacaagtat ttaatatctt ttttaagtga gtaaaaatat tctagctatt ggcaaataga
     2401 ttacagtaga caattagaga atatgatgag gttgactctt ccaatcaatc agtatcaagc
     2461 taagtcgact gtgtgcgtgt tggtttgttt tccttttttt tcttttttt tactataaca
     2521 gtcttattta aattcggattt ttgtattcag tttgtatgac agtagactag gccagcctag
     2581 ctgtaaatgc atgctacact ttcctctccg tcatgcagct aagcttgctt gtcagcccgt
     2641 gcttaacttg tggtgaactg gactttttgtt taaaataagc aaaactgtaa gggacagtgc
     2701 atcggccttt ccccgtcttc ttaagtagaa gtttgctcat ccttccgaat tcgagggcgt
     2761 gttatccatc tgatgcttgc tgttaacccc agcacttggc agcttcccat ccatcctgtg
     2821 ttactatgg ttccatctca gcacacagtg tatagcctat tagcaataac ttcctatgta
     2881 tagagaacat tttcatctta ctgagccagc ccatttgcct aatgactata cagacagtct
     2941 aaattgcaat atatctgtaa atgtgtatat actataaata cacaatatac ttacattttt
     3001 ctgggccaaa cagcattttcc tttcccctcc ttttcctctc ccctctcttg ccttatgatg
     3061 aatttgtttg agggcatttt cttcatgaac aaggcttgag atgcatattc ctaggtttga
     3121 ttttgttttt gttttttttc tttctgtgaa atgggtatta ttatccccaa ggagagctgc
     3181 gcagtgagag ccagtctagt tatgacccac tacgtgttgt tttttacatt actctttgat
     3241 gagtcgcgtt ttgctgagca cgtttgttgt ccgtcgtgaa atgttagtgga gtttggtgga
     3301 tgccctttgc actgctgaag tgaaatcttg ctgactactt agcttttttg ttgatctcag
     3361 tatgggcaaa gaaacaaaat ccaaatggct ggacagactt ctgtaaatat tttttaaaa
     3421 aaacaaaaca aatgttctgt atgtgtgtgt tttacgtggt catgaagtat ttccttaaag
```

```
3481 cgaagctttg agtacacaga gagtaactct ttgtttagca agtttggagt gaagataaaa
3541 tagcaagaaa tactgtaaaa tgtcaaagca ccacaccaga actgtaggag tccagattta
3601 taataaactg tctttttaaa aagaaaacaa ttaagtgtct agaatatatc atgttttatt
3661 atttaaaatc attgtcttaa gttttttgtt gagagaataa aaaatttgaa tacaatcaa
```

8.1.3 Human Psap

```
LOCUS       HUMSAPABCD   2171 bp    mRNA         PRI      09-JAN-1995
DEFINITION  Human saposin proteins A-D mRNA, complete cds.
ACCESSION   M32221
VERSION     M32221.1  GI:337761
KEYWORDS    saposin.
SOURCE      Human lymphoblast, cDNA to mRNA.
  ORGANISM  Homo sapiens
            Eukaryota; Metazoa; Chordata; Craniata; Vertebrata; Euteleostomi;
            Mammalia; Eutheria; Primates; Catarrhini; Hominidae; Homo.
REFERENCE   1  (bases 1 to 2171)
  AUTHORS   Kretz,K.A., Carson,G.S., Morimoto,S., Kishimoto,Y., Fluharty,A.L.
            and O'Brien,J.S.
  TITLE     Characterization of a mutation in a family with saposin B
            deficiency: a glycosylation site defect
  JOURNAL   Proc. Natl. Acad. Sci. U.S.A. 87 (7), 2541-2544 (1990)
  MEDLINE   90207231
COMMENT     Draft entry and computer-readable sequence for [1] kindly submitted
            by J.S.O'Brien, 22-MAR-1990.
FEATURES             Location/Qualifiers
     source          1..2171
                     /organism="Homo sapiens"
                     /db_xref="taxon:9606"
                     /map="10q21-q22"
     gene            7..1581
                     /gene="PSAP"
     CDS             7..1581
                     /gene="PSAP"
                     /note="prosaposin"
                     /codon_start=1
                     /protein_id="AAA60303.1"
                     /db_xref="GI:337762"
                     /db_xref="GDB:G00-120-366"
                     /translation="MYALFLLASLLGAALAGPVLGLKECTRGSAVWCQNVKTASDCGA
                     VKHCLQTVWNKPTVKSLPCDICKDVVTAAGDMLKDNATEEEILVYLEKTCDWLPKPNM
                     SASCKEIVDSYLPVILDIIKGEMSRPGEVCSALNLCESLQKHLAELNHQKQLESNKIP
                     ELDMTEVVAPFMANIPLLLYPQDGPRSKPQPKDNGDVCQDCIQMVTDIQTAVRTNSTF
                     VQALVEHVKEECDRLGPGMADICKNYISQYSEIAIQMMMHMQPKEICALVGFCDEVKE
                     MPMQTLVPAKVASKNVIPALELVEPIKKHEVPAKSDVYCEVCEFLVKEVTKLIDNNKT
                     EKEILDAFDKMCSKLPKSLSEECQEVVDTYGSSILSILLEEVSPELVCSMLHLCSGTR
                     LPALTVHVTQPKDGGFCEVCKKLVGYLDRNLEKNSTKQEILAALEKGCSFLPDPYQKQ
                     CDQFVAEYEPVLIEILVEVMDPSFVCLKIGACPSAHKPLLGTEKCIWGPSYWCQNTET
                     AAQCNAVEHCKRHVWN"
     mat_peptide     184..435
                     /gene="PSAP"
                     /note="saposin A"
     mat_peptide     589..831
                     /gene="PSAP"
                     /note="saposin B"
     variation       656
                     /gene="PSAP"
                     /note="c in wt; t in saposin B deficiency (Thr->Ile)"
     mat_peptide     937..1176
                     /gene="PSAP"
                     /note="saposin C"
     mat_peptide     1219..1467
                     /gene="PSAP"
                     /note="saposin D"
BASE COUNT      496 a    565 c    610 g    500 t
ORIGIN      Chromosome 10, q21-q22.
        1 cgcgctatgt acgccctctt cctcctggcc agcctcctgg gcgcggctct agccggcccg
```

```
  61 gtccttggac tgaaagaatg caccaggggc tcggcagtgt ggtgccagaa tgtgaagacg
 121 gcgtccgact gcggggcagt gaagcactgc ctgcagaccg tttggaacaa gccaacagtg
 181 aaatcccttc cctgcgacat atgcaaagac gttgtcaccg cagctggtga tatgctgaag
 241 gacaatgcca ctgaggagga gatccttgtt tacttggaga agacctgtga ctggcttccg
 301 aaaccgaaca tgtctgcttc atgcaaggag atagtggact cctacctccc tgtcatcctg
 361 gacatcatta aaggagaaat gagccgtcct ggggaggtgt gctctgctct caacctctgc
 421 gagtctctcc agaagcacct agcagagctg aatcaccaga agcagctgga gtccaataag
 481 atcccagagc tggacatgac tgaggtggtg gcccccttca tggccaacat ccctctcctc
 541 ctctaccctc aggacggccc ccgcagcaag ccccagccaa aggataatgg ggacgtttgc
 601 caggactgca ttcagatggt gactgacatc cagactgctg tacggaccaa ctccaccttt
 661 gtccaggcct tggtggaaca tgtcaaggag gagtgtgacc gcctgggccc tggcatggcc
 721 gacatatgca agaactatat cagccagtat tctgaaattg ctatccagat gatgatgcac
 781 atgcaaccca agagatctg tgcgctggtt gggttctgtg atgaggtgaa agagatgccc
 841 atgcagactc tggtccccgc caaagtggcc tccaagaatg tcatccctgc cctggaactg
 901 gtgtgagccca ttaagaagca cgaggtccca gcaaagtctg atgtttactg tgaggtgtgt
 961 gaattcctgg tgaaggaggt gaccaagctg attgacaaca acaagactga gaaagaaata
1021 ctcgacgctt ttgacaaaat gtgctcgaag ctgccgaagt ccctgtcgga agagtgccag
1081 gaggtggtgg acacgtacgg cagctccatc ctgtccatcc tgctggagga ggtcagccct
1141 gagctggtgt gcagcatgct gcacctctgc tctggcacgc ggctgcctgc actgaccggt
1201 cacgtgactc agccaaagga cggtggcttc tgcgaagtgt gcaagaagct ggtgggttat
1261 ttggatcgca acctggagaa aaacagcacc aagcaggaga tcctggctgc tcttgagaaa
1321 ggctgcagct tcctgccaga cccttaccag aagcagtgtg atcagtttgg ggcagagtac
1381 gagcccgtgc tgatcgagat cctggtggag gtgatggatc cttccttcgt gtgcttgaaa
1441 attggagcct gcccctcggc ccataagccc ttgttgggaa ctgagaagtg tatatggggc
1501 ccaagctact ggtgccagaa cacagagca gcagcccagt gcaatgctgt cgagcattgc
1561 aaacgccatg tgtggaacta ggaggaggaa tattccatct tggcagaaac cacagcattg
1621 gtttttttct acttgtgtgt ctggggggaat gaacgcacag atctgtttga ctttgttata
1681 aaaatagggc tcccccacct cccccatttc tgtgtccttt attgtagcat tgctgtctgc
1741 aagggagccc ctagccccctg gcagacatag ctgcttcagt gcccctttc tctctgctag
1801 atggatgttg atgcactgga ggtctttag cctgcccttg catggcgcct gctggaggag
1861 gagagagctc tgctggcatg agccacagtt tcttgactgg aggccatcaa ccctcttggt
1921 tgaggccttg ttctgagccc tgacatgtgc ttgggcacctg gtgggcctgg gcttctgagg
1981 tggcctcctg ccctgatcag ggaccctccc cgctttcctg ggcctctcag ttgaaccaaa
2041 gcagcaaaac aaaggcagtt ttatatgaaa gattagaagc ctggaataat caggcttttt
2101 aaatgatgta attcccactg taatagcata gggattttgg aagcagctgc tggtggcttg
2161 ggacatcagt g
```

8.1.4 Mouse pSap

```
LOCUS       NM_001146122            2664 bp    mRNA    linear   ROD 21-MAR-2010
DEFINITION  Mus musculus prosaposin (Psap), transcript variant 4, mRNA.
ACCESSION   NM_001146122
VERSION     NM_001146122.1  GI:225735652
KEYWORDS    .
SOURCE      Mus musculus (house mouse)
  ORGANISM  Mus musculus
            Eukaryota; Metazoa; Chordata; Craniata; Vertebrata; Euteleostomi;
            Mammalia; Eutheria; Euarchontoglires; Glires; Rodentia;
            Sciurognathi; Muroidea; Muridae; Murinae; Mus.
REFERENCE   10 (bases 1 to 2664)
  AUTHORS   Tsuda,M., Sakiyama,T., Endo,H. and Kitagawa,T.
  TITLE     The primary structure of mouse saposin
  JOURNAL   Biochem. Biophys. Res. Commun. 184 (3), 1266-1272 (1992)
  PUBMED    1590788
COMMENT     VALIDATED REFSEQ: This record has undergone validation or
            preliminary review. The reference sequence was derived from
            BY325400.1, AK159221.1 and S36200.1.

            Transcript Variant: This variant (4) lacks an exon in the
            coding
            region and uses an alternate in-frame splice site, compared to
            variant 2, resulting in a shorter protein (isoform D), compared
            to isoform B.

            Publication Note:  This RefSeq record includes a subset of the
            publications that are available for this gene. Please see the
            Entrez Gene record to access additional publications.
```

```
             COMPLETENESS: complete on the 3' end.
PRIMARY      REFSEQ_SPAN            PRIMARY_IDENTIFIER PRIMARY_SPAN        COMP
             1-88                   BY325400.1         1-88
             89-2642                AK159221.1         7-2560
             2643-2664              S36200.1           2546-2567
FEATURES             Location/Qualifiers
     source          1..2664
                     /organism="Mus musculus"
                     /mol_type="mRNA"
                     /strain="C57BL/6"
                     /db_xref="taxon: 10090"
                     /chromosome="10"
                     /map="10 35.0 cM"
     gene            1..2664
                     /gene="Psap"
                     /gene_synonym="AI037048; SGP-1"
                     /note="prosaposin"
                     /db_xref="GeneID:19156"
                     /db_xref="MGI:97783"
     CDS             116..1777
                     /gene="Psap"
                     /gene_synonym="AI037048; SGP-1"
                     /note="isoform D preproprotein is encoded by
                     transcript
                     variant 4; snoRNA MBII-198; sulfated glycoprotein 1"
                     /codon_start=1
                     /product="sulfated glycoprotein 1 isoform D
                     preproprotein"
                     /protein_id="NP_001139594.1"
                     /db_xref="GI:225735653"
                     /db_xref="GeneID:19156"
                     /db_xref="MGI:97783"
                     /translation="MYALALFASLLATALTSPVQDPKTCSGGSAVLCRDVKTAVDCGA
                     VKHCQQMVWSKPTAKSLPCDICKTVVTEAGNLLKDNATQEEILHYLEKTCEWIHDSSL
                     SASCKEVVDSYLPVILDMIKGEMSNPGEVCSALNLCQSLQEYLAEQNQKQLESNKIPE
                     VDMARVVAPFMSNIPLLLYPQDHPRSQPQPKANEDVCQDCMKLVSDVQTAVKTNSSFI
                     QGFVDHVKEDCDRLGPGVSDICKNYVDQYSEVCVQMLMHMQPKEICVLAGFCNEVKRV
                     PMKTLVPATETIKNILPALEMMDPYENLVQAHNVILCQTCQFVMNKFSELIVNNATEE
                     LLVKGLSNACALLPDPARTKCQEVVGTFGPSLLDIFIHEVNPSSLCGVIGLCAARPEL
                     VEALEQPAPAIVSALLKEPTPPKQPAQPKQSALPAHVPPQKNGGFCEVCKKLVLYLEH
                     NLEKNSTKEEILAALEKGCSFLPDPYQKQCDDFVAEYEPLLLEILVEVMDPGFVCSKI
                     GVCPSAYKLLLGTEKCVWGPSYWCQNMETAARCNAVDHCKRHVWN"
ORIGIN
        1 gtttgggcgg ggcccgcgcc tgcgcactgg cgtctgcggc gaggttgggg ggtagattta
       61 tattcgtctg gtcagctgac gttctgcatt gcagcctgcg aagtgaagcg gcgccatgta
      121 cgccctcgcc ctcttcgcca gccttctggc caccgctctg accagccctg tccaagaccc
      181 gaagacatgc tctggggct cagcagtgct gtgcagagat gtgaagacgg cggtggactg
      241 tggggccgtg aagcactgcc agcagatggt ctggagcaag cccacagcga aatccctcc
      301 ttgcgacata tgcaaaactg ttgtcaccga agctgggaac ttgctgaaag ataatgctac
      361 gcaggaggag atccttcatt acctggagaa gacctgtgag tggattcatg actccagcct
      421 gtcggcctcg tgcaaggagg tggttgactc ttacctgcct gtcatcctgg acatgattaa
      481 gggcgagatg agcaaccctg gggaagtgtg ctctgcgctc aacctctgcc agtcccttca
      541 ggagtacttg gccagacaaa accagaaaca gcttgagtcc aacaagatcc cggaggtgga
      601 catggcccgt gtggttgccc ccttcatgtc caacatccct ctcctgctgt accctcagga
      661 tcaccccgc agccagcccc aacctaaggc taacgaggac gtctgccagg actgtatgaa
      721 gctggtgtct gatgtccaga ctgctgtgaa gaccaactcc agctttatcc aggttcgt
      781 ggaccacgtg aaggaggatt gtgaccgctt ggggccaggc gtgtctgaca tatgcaagaa
      841 ctacgtggac cagtattccg aggtctgtgt ccagatgttg atgcacatgc aacccaagga
      901 aatctgtgtg ctggctggct tctgtaatga ggtcaagaga gtgccaatga agactctggt
      961 ccctgccacc gagaccatta agaacatcct ccctgccctg gagatgatgg acccctaga
     1021 gaatctggtc caggcccaca atgtgatttt atgccagacc tgtcagtttg tgataataa
     1081 gttttctgag ctgattgtca ataatgccac tgaggagctc ctagttaaag gtttgagcaa
     1141 cgcatgcgca ctgctccccg atcctgccag aaccaagtgc caggaggtgg tgggaacatt
     1201 tggcccctcc tcgttggaca tctttatcca tgaggtaaac cccagctcct tgtgcggtgt
     1261 gatcggcctc tgtgctgccc gcccggagtt ggtggaggca cttgagcagc ctgcgccagc
     1321 cattgtatct gcactgctca aagagcccac accgccaaag cagcccgcac ctccaaagca
     1381 gtcggcattg cccgcccatg tgcctcctca aagaaatggt gggttctgtg aggtgtgcaa
     1441 gaaactggtc ctctatttgg aacataacct ggagaaaaac agcaccaagg aggaaatcct
     1501 ggccgcactt gagaagggct gcagcttcct gccagaccct taccagaagc agtgcgatga
```

```
1561 ctttgtggct gagtatgagc ccttgctatt ggagatcctc gtggaagtga tggatcctgg
1621 atttgtgtgc tcgaaaattg gagtttgccc ttctgcctat aagctgctgc tgggaaccga
1681 gaagtgtgtc tggggcccta gctactggtg tcagaacatg gagactgccg cccgatgcaa
1741 tgctgtcgat cattgcaaac gccatgtgtg gaactagttt cccagctgca gaagtcacct
1801 acttgtgggt ctagggtaat gaacacatag atctatttga cttaataagt aggaaccccc
1861 tttgcccttc ccccatctcc tctcccttac tgtagcattt ctgtcatgta agaggtgctg
1921 acagccactt ccgtgtcccc tttctgctcg aaggatgagg ataccttggg catcagctcc
1981 ccggctgccc ttttcaccca cctgctggag ggggtggtg agccagaggg caggagcatt
2041 ttctgagccc tttcttggtg tgtgggggat ctatggccat ctcctaccat gagggagcta
2101 cccagcttcc tgtggtacca aggagttatt ttggatgatt agaagcacag aatgatcagg
2161 cctttagagc gatggaatgg ccattgtcat agcacagaga tttcagaagc acctgcaggt
2221 ggcttgcttg ggatgttgct gtccctgggt cagccttcca ttctgctttc ctgtcttccc
2281 gtctgccttg ttggggttct gtggggtagg gtggggaggg gaaacttgtg aatgtaactt
2341 gcctgtgccg tgtgacggtc acgtgggcct ggtcttttgt gtgtgaggcc cttgaccgtg
2401 tggcctctgc ctggctgttt ggggtcctgc acggctttcc caccacctgt agctcttgtt
2461 gacctgcctg ttcacctcat gagtgaagcg tctgcctggc agtgggccat gaactgaggg
2521 gtctctgtgt agagtagaag cttcctgtgc ctccggttgc caggagacag cctgtgcagt
2581 taaatggacc tagattttgt tttgcactaa agtttctgtg acttaataaa gttctgttaa
2641 ccaacagaaa aaaaaaaaaa aaaa
//
```

8.2 Plasmids

8.2.1 pBluesscript SKII(-)

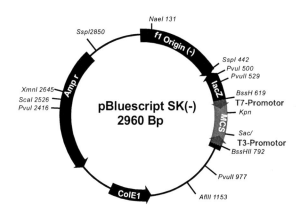

8.2.2 pcDNA3.1HisA

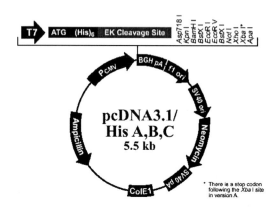

5' end of hCMV promoter/enhancer
↓

```
209   CGATGTACGG GCCAGATATA CGCGTTGACA TTGATTATTG ACTAGTTATT AATAGTAATC AATTACGGGG
```

```
                                    enhancer region (5' end)
279   TCATTAGTTC ATAGCCCATA TATGGAGTTC CGCGTTACAT AACTTACGGT AAATGGCCCG CCTGGCTGAC
```

```
349   CGCCCAACGA CCCCCGCCCA TTGACGTCAA TAATGACGTA TGTTCCCATA GTAACGCCAA TAGGGACTTT
```

```
419   CCATTGACGT CAATGGGTGG ACTATTTACG GTAAACTGCC CACTTGGCAG TACATCAAGT GTATCATATG
```

```
489   CCAAGTACGC CCCCTATTGA CGTCAATGAC GGTAAATGGC CCGCCTGGCA TTATGCCCAG TACATGACCT
```

```
559   TATGGGACTT TCCTACTTGG CAGTACATCT ACGTATTAGT CATCGCTATT ACCATGGTGA TGCGGTTTTG
```

```
629   GCAGTACATC AATGGGCGTG GATAGCGGTT TGACTCACGG GGATTTCCAA GTCTCCACCC CATTGACGTC
```

```
      enhancer region (3' end)
699   AATGGGAGTT TGTTTTGGCA CCAAAATCAA CGGGACTTTC CAAAATGTCG TAACAACTCC GCCCCATTGA
```

```
      CAAT                                         TATA     3' end of hCMV    putative transcriptional start
769   CGCAAATGGG CGGTAGGCGT GTACGGTGGG AGGTCTATAT AAGCAGAGCT CTCTGGCTAA CTAGAGAACC
```

```
                                    T7 promoter priming site
839   CACTGCTTAC TGGCTTATCG AAATTAATAC GACTCACTAT AGGGAGACCC AAGCTGGCTA GCGTTTAAAC
```

```
      Hind III                              ProBond binding domain
909   TTAAGCTTAC C ATG GGG GGT TCT CAT CAT CAT CAT CAT CAT GGT ATG GCT AGC ATG ACT
                    Met Gly Gly Ser His His His His His His Gly Met Ala Ser Met Thr
```

```
                         Anti-Xpress™ Antibody Epitope          Asp718 I Kpn I  Bam H I
968   GGT GGA CAG CAA ATG GGT CGG GAT CTG TAC GAC GAT GAC GAT AAG GTA CCT AGG ATC CAG
      Gly Gly Gln Gln Met Gly Arg Asp Leu Tyr Asp Asp Asp Asp Lys Val Pro Arg Ile Gln
                                                    Enterokinase recognition sequence    Enterokinase cleavage site
```

```
      Bst X I  Eco R I        Eco R V          Bst XI  Not I   Xho I     Xba I        Apa I
1028  TGT GGT GGA ATT CTG CAG ATA TCC AGC ACA GTG GCG GCC GCT CGA GTC TAG AGGGCCCGTT
      Cys Gly Gly Ile Leu Gln Ile Ser Ser Thr Val Ala Ala Ala Arg Val ***
```

```
      pcDNA3.1/BGH reverse priming site
1089  TAAACCCGCT GATCAGCCTC GACTGTGCCT TCTAGTTGCC AGCCATCTGT TGTTTGCCCC TCCCCCGTGC
```

```
                         BGH poly (A) site
1155  CTTCCTTGAC CCTGGAAGGT GCCACTCCCA CTGTCCTTTC CTAATAAAAT GAGGAAATTG CATCGCATTG
```

```
1229  TCTGAGTAGG TGTCATTCTA TTCTGGGGGG TGGGGTGGGG CAGGACAGCA AGGGGGAGGA TTGGGAAGAC
```

8.3. List of publications.

Arenz C, Cryns A, Diallo M, Sandhoff K, Schepers U (2003) Analysis of skin glycosphingolipid metabolism by RNA interference in transgenic mice. European Journal of Cell Biology 82: 97-97

Diallo M, Arenz C, Schmitz K, Sandhoff K, Schepers U (2003a) Long endogenous dsRNAs can induce complete gene silencing in mammalian cells and primary cultures. Oligonucleotides 13: 381-92

Diallo M, Arenz C, Schmitz K, Sandhoff K, Schepers U (2003b) RNA interference: analyzing the function of glycoproteins and glycosylating proteins in mammalian cells. Methods Enzymol 363: 173-90

Diallo M, Arenz C, Schmitz K, Sandhoff K, Schepers U (2003c) RNA interference: Analyzing the function of glycoproteins and glycosylating proteins in mammalian cells. Recognition of Carbohydrates in Biological Systems, Part B: Specific Applications 363: 173-190

Diallo M, Arenz C, Schmitz K, Sandhoff K, Schepers U (2003d) Stable post transcriptional gene silencing of glucosylceramicle-synthase by RNA interference in human cells. European Journal of Cell Biology 82: 98-98

Diallo M, Schepers U (2003) RNA Interference: Everything What One Ever Wanted to Know About Silence! Bioforum Europe 3: 143-145

Diallo M, Schmitz K, Schepers U (2005) RNA interference: RNAid for future therapeutics. Humana Press, Totowa, NY

Schmitz K, Arenz C, Diallo M, Cryns A, Hoffman S, Sandhoff K, Schepers U (2003a) RNA interference: New approaches to study protein function in mammalian cells and organisms. European Journal of Cell Biology 82: 99-100

Schmitz K, Diallo M, Arenz C, Mundegar R, Sandhoff K, Schepers U (2003b) Synthesis of a new carrier-system for transduction of siRNAs into cells for the local transient posttranscriptional gene silencing in mammals. European Journal of Cell Biology 82: 100-100

Walter J, Prager K, Diallo M, Wahle T, Lütjohann D, Haass C, Sandhoff K, Schepers U, Tamboli I (2004) Glycosphingolipids regulate subcellular transport of the beta-amyloid procursor protein and the generation of amyloid beta-peptide. Neurobiology of Aging 25: S38-S38

9. References

Abe, A., Radin, N. S. and Shayman, J. A. (1996) Biochim. Biophys. Acta 1299, 333–341

Acharya U and Acharya JK (2005) Enzymes of sphingolipid metabolism in Drosophila melanogaster. Cell Mol Life Sci 62: 128-142.

Ahlquist P (2002). "RNA-dependent RNA polymerases, viruses, and RNA silencing". Science 296 (5571): 1270–1273.

Ahn VE, Faull KF, Whitelegge JP, Fluharty AL, Prive GG. 2003. Crystal structure of saposin B reveals a dimeric shell for lipid binding. Proc. Natl. Acad. Sci. USA 100:38–43

Ambros, V., B. Bartel, et al. (2003) A uniform system for microRNA annotation Rna 9(3): 277-9

Aravind L., Hidemi Watanabe, David J. Lipman, and Eugene V. Koonin (2000). "Lineage-specific loss and divergence of functionally linked genes in eukaryotes". *Proceedings of the National Academy of Sciences* 97 (21): 11319–11324.

ARENZ, C., and Schepers U. (2003) RNA interference An ancient mechanismor a state of the art therapeutic application. Naturwissenschaften 90: 345-359

Arrese EL, Wells MA (1997) Adipokinetic hormone induced lipolysis in the fat body of insect, Meduca sexta: synthesis of sn-1,2-diacylglycerols. J Lipid Res 38: 68-76.

Bagasra O, Prilliman KR (2004). RNA interference *J. Mol. Histol.* 35 (6): 545–553.

Bagga S., Bracht J., Hunter S., Massirer K., Holtz J., Eachus R., and Pasquinelli AE. Regulation by let-7 and lin-4 miRNAs results in mRNA target degradation. Cell 2005; 122: 553-563.

Bartel DP, microRNAs: genomics, biogenesis, mechanism and function. Cell 2004; 116: 281-297

Bauer R, Voelzmann A, Breiden B, Schepers U, Farwanah H, Hahn I, Eckardt F, Sandhoff K, and Hoch M (2009). Schlank, a member of the dihydroceramide synthase family controls growth and body fat in Drosophila. EMBO Journal (2009); 28: 3706-3716.

Bennett, CF Chiang, MY, Chan H, Shoemaker JE, and Mirabelli CK (1992) Cationic lipids enhance cellular uptake and activity of phosphorothiotat antisense oligionucleotides Mol Pharmaco 41:1023-1033

Bennett CF, Mirejovsky D, Crooke RM, Tai YJ, Felgner J, Sridhar CN, Wheeler CJ, and Felgner PL (1998) Structural requirements for cationic lipid mediated phosphorothiotat oligionucleotides delivery to cells in culture J Drug target 5:1023-1033

Bennett, M.K.; Lopez, J.M.; Sanchez, H.B and Osborn, T.F., 1995. J. Biol. Chem. 270 25578-25583

Berent SL, Radin NS (1981) beta-Glucosidase activator protein from bovine spleen ("coglucosidase"). Arch Biochem Biophys 208: 248-60

Berent SL, Radin NS. 1981. Mechanism of activation of glucocerebrosidase by co-β-glucosidase (glucosidase activator protein). *Biochim. Biophys. Acta* 664:572–82

Berkhout, B; ter Brake, O (2010). "RNAi Gene Therapy to Control HIV-1 Infection". *RNA Interference and Viruses: Current Innovations and Future Trends*. Caister Academic Press. ISBN 978-1-904455-56-1.

Bernstein E and Caudy A (2001) Role for a bidentate ribonuclease in the initiation step of RNA interference Nature 409(6818): 363-366

Bernstein E, and Denli A M (2001) The rest is silence [Review] RNAi Publication of the RNA Society 7(11): 1509-1521

Blaszczyk J, and Tropea J E (2001) Crystallographic and modeling studies of RNase III suggest a mechanism for double-stranded RNA cleavage Structure 9(12): 1225-1236

Bosher J M, and Labouesse M (2000) RNA interference: genetic wand and genetic watchdog Nat Cell Biol 2(2): E31-6

Boutros M, Kiger A, Armknecht S, Kerr K, Hild M, Koch B, Haas S, Paro R, Perrimon N (2004). "Genome-wide RNAi analysis of growth and viability in Drosophila cells". *Science* **303** (5659): 832–5.

Bradova V, Smid F, Ulrich-Bott B, Roggendorf W, Paton BC, Harzer K (1993) Prosaposin deficiency: further characterization of the sphingolipid activator protein-deficient sibs. Multiple glycolipid elevations (including lactosylceramidosis), partial enzyme deficiencies and ultrastructure of the skin in this generalized sphingolipid storage disease. Hum Genet 92: 143-52

Brennecke J, Hipfner DR, Stark A, Russell RB, and Cohen SM (2003) b*antam* encodes a developmentally regulated microRNA that controls cell proliferation and regulates the proapoptotic gene *hid in Drosophila Cell* **113**:25-26

Brown DA, London E (June 2000). "Structure and function of sphingolipid- and cholesterol-rich membrane rafts". *J. Biol. Chem.* **275** (23): 17221–4.

Brown and Goldstein, 1997 Cell 89, 331-340

Brummelkamp T R, Agami R (2002) A system for stable expression of short interfering RNAs in mammalian cells Science **296**(5567): 550-553.

Burger, K. N. J., van der Bijl, P. and van Meer, G. (1996) J. Cell Biol. 133, 15–28

Burkhardt JK, Hüttler S, Klein A, Möbius W, Habermann A, et al. 1997. Accumulation of sphingolipids in SAP-precursor (prosaposin)-deficient fibroblasts occurs as intralysosomal membrane structures and can be completely reversed by treatment with human SAP-precursor. *Eur. J. Cell Biol.* 73:10–18

Cabot R A and Prather R S (2003) Cleavage stage porcine embryos may have differing developmental requirements for karyopherins alpha2 and alpha3 Mol Reprod Dev **64**(3): 292-301

Calin GA, Dumitru C D, Shimizu M, Bichi R, Zupo S, Noch E, Adler H, Rattan S, Keating M, Rai K, Rassenti L. Kipps T, Negrini M, Bullrich F, and Croce CM (2002) Frequent deletions and down-regulation of microRNA genes miR15 and miR16 at 13q14 in chronic lymphocytic leukemia Proc. Natl. Acad. Sci. **99**:15524-15529

Caplen NJ, Parrish S, Imani F, Fire A, and Morgan RA (2001) Specific inhibition of gene expression by small double-stranded RNAs in invertebrate and vertebrate systems Pro natl. Acad Sci **98**: 9742-9747

Caplen NJ, Parrish S, Imani F, Fire A. (2000). dsRNA-mediated gene silencing in cultured *Drosophila* cells: a tissue culture model for the analysis of RNAi. *Gene* Vol., 252; pp, 95-105

Campana WM, Hiraiwa M, O'Brien JS (1998) Prosaptide activates the MAPK pathway by a G-protein-dependent mechanism essential for enhanced sulfatide synthesis by Schwann cells. Faseb J 12: 307-14

Castanotto, Daniela; Rossi, John J. (22 January 2009). "The promises and pitfalls of RNA-interference-based therapeutics". *Nature* **457** (7228): 426–433.

Causeret et al. "Further characterization of rat dihydroceramide desaturase: tissue distribution, subcellular localization, and substrate specificity". Lipids. (2005) 35:1117-1125.

Chatterjee S, Großhans H (September 2009). "Active turnover modulates mature microRNA in *Caenorhabditis elegans*". *Nature* **461** (7263): 546–459.

Christomanou H, Aignesberger A, Linke RP. 1986. Immunochemical characterization of two activator proteins stimulating enzymic sphingomyelin degradation in vitro. Absence of one of them in a human Gaucher disease variant. *Biol. Chem. Hoppe-Seyler* 367:879–90

Ciaffoni F, Tatti M, Salvioli R, Vaccaro AM (2003) Interaction of saposin D with membranes: effect of anionic phospholipids and sphingolipids. Biochem J 373: 785-792.

Ciaffoni F, Salvioli R, Tatti M, Arancia G, Crateri P, Vaccaro AM. 2001. Saposin D solubilizes anionic phospholipid-containing membranes. *J. Biol. Chem.* 276:31583–89

Cioca DP, Aoki Y, and Kiyosawa K (2003) RNAi is a functional pathway with therapeutic potential in human myeloid leukemia cell lines Cancer Gene Ther **10**:125-133

Clemens J.C., Worby C.A., Simmonson Leff, N. (2000): Use of dsRNA RNA interference in *Drosophila* cell lines to dissect signal transduction pathways, *Proc. Natl. Acad. Sci,* USA, Vol. 97, pp. 6499-6503

Coetzee, T. et al. (1996) Cell 86, 209–219

Coste, H., Martel, M. B. and Got, R. (1986) Biochim. Biophys. Acta 858, 6–12

Crowe S (2003). "Suppression of chemokine receptor expression by RNA interference allows for inhibition of HIV-1 replication, by Martínez et al.". *AIDS* **17 Suppl 4**: S103–5.

Cullen L, Arndt G (2005). "Genome-wide screening for gene function using RNAi in mammalian cells". *Immunol Cell Biol* **83** (3): 217–23.

Daneholt, Bertil. Advanced Information: RNA interference. *The Nobel Prize in Physiology or Medicine 2006.* Retrieved 2007-01-25.

DaRocha W, Otsu K, Teixeira S, Donelson J (2004). "Tests of cytoplasmic RNA interference (RNAi) and construction of a tetracycline-inducible T7 promoter system in Trypanosoma cruzi". *Mol Biochem Parasitol* **133** (2): 175–86.

Diallo, M., Arenz, C., Schmitz, K., Sandhoff, K. & Schepers, U. RNA Interference: A new method to analyze the function of glycoproteins and glycosylating proteins in mammalian cells.Knockout experiments with UDP-glucose/ ceramide-glucosyltransferase. Methods in Enzymology 363, 173-190 (2003).

Donze, O. and Picard, D. (2002), 'RNAinterference in mammalian cells using siRNAs synthesized with T7 RNA polymerase', Nucleic Acids Res., Vol. 30, p.46.

Drinnenberg IA, Weinberg DE, Xie KT, Nower JP, Wolfe KH, Fink GR, Bartel DP (2009). "RNAi in Budding Yeast". *Science.*

Eckert D G. and Bass B (2006). "RDE-4 preferentially binds long dsRNA and its dimerization is necessary for cleavage of dsRNA to siRNA". *RNA* **12** (5): 807–818.

Elbashir SM, Lendeckel W and Tuschl T 2001 RNA interference is mediated by 22 - 23-nucleotides RNAs *Genes and Development* 15 188-200

Elbashir, S. M., Harborth, J., Lendeckel, W. (2001), 'Duplexes of 21-nucleotide RNAs mediate RNA interference in culturedmammalian cells', Nature, Vol. 411, pp. 494-498.

Ericsson, J, 1986. Proc. Natl. Acad. Sci. USA. 93 945-950

Fabbro D, Grabowski GA. 1991. Human acid β-glucosidase. Use of inhibitory and activating monoclonal antibodies to investigate the enzyme's catalytic mechanism and saposin A and C binding sites. *J. Biol. Chem.* 266:15021–27

Farwanah H, Nuhn P, Neubert R, Raith K (2003) Normal phase LC separation of stratum corneum ceramides with detection by evaporative light scattering and APCI mass spectrometry. Anal. Chim. Acta 492: 233-239.

Fire A, Xu S, Montgomery M, Kostas S, Driver S, Mello C (1998). "Potent and specific genetic interference by double-stranded RNA in *Caenorhabditis elegans*". *Nature* **391** (6669): 806–811.

Fischer G, Jatzkewitz H. 1977. The activator of cerebroside sulphatase. Binding studies with enzyme and substrate demonstrating the detergent function of the activator protein. *Biochim. Biophys. Acta* 481:561–72

Fujibayashi S, Wenger DA (1986a) Biosynthesis of the sulfatide/GM1 activator protein (SAP-1) in control and mutant cultured skin fibroblasts. Biochim Biophys Acta 875: 554-62

Fujibayashi S, Wenger DA (1986b) Synthesis and processing of sphingolipid activator protein-2 (SAP-2) in cultured human fibroblasts. J Biol Chem 261: 15339-43

Fujita N, Suzuki K, Vanier MT, Popko B, Maeda N, et al. 1996. Targeted disruption of the mouse sphingolipid activator protein gene: a complex phenotype, including severe leukodystrophy and wide-spread storage of multiple sphingolipids. *Hum. Mol. Genet.* 5:711–25

Futerman AH (December 2006). "Intracellular trafficking of sphingolipids: relationship to biosynthesis". Biochim. Biophys. Acta 1758 (12): 1885–1892.

Futerman, A. H. and Pagano, R. E. (1991) Biochem. J. 280, 295–302

Fyrst H., Herr D. R., Harris G. L.and Saba J. D., Characterization of free endogenous C14 and C16 sphingoid bases from Drosophila melanogaster. Journal of Lipid Research 45, (2004) 54-62.

Ge G, Wong G, Luo B (2005). "Prediction of siRNA knockdown efficiency using artificial neural network models". *Biochem Biophys Res Commun* 336 (2): 723–8.

Geldhof P, Murray L, Couthier A, Gilleard J, McLauchlan G, Knox D, Britton C (2006). "Testing the efficacy of RNA interference in Haemonchus contortus". *Int J Parasitol* 36 (7): 801–10.

Geldhof P, Visser A, Clark D, Saunders G, Britton C, Gilleard J, Berriman M, Knox D. (2007). "RNA interference in parasitic helminths: current situation, potential pitfalls and future prospects". *Parasitology* 134: 1–11.

Gibbons GF, Islam K, Pease RJ (2000) Mobilisation of triacylglycerol stores. *Biochim Biophys Acta* 1483: 37-57.

Goni FM, Alonso A. 2002. Sphingomyelinases: enzymology and membrane activity. *FEBS Lett.* 531:38–46

Gregory R, Chendrimada T, Shiekhattar R (2006). "MicroRNA biogenesis: isolation and characterization of the microprocessor complex". *Methods Mol Biol* 342: 33–47.

Gregory R, Chendrimada T, Cooch N, Shiekhattar R (2005). "Human RISC couples microRNA biogenesis and posttranscriptional gene silencing". *Cell* 123 (4): 631–640.

Grimm D, Streetz K, Jopling C, Storm T, Pandey K, Davis C, Marion P, Salazar F, Kay M (2006). "Fatality in mice due to oversaturation of cellular microRNA/short hairpin RNA pathways". *Nature* 441 (7092): 537–41.

Hahn SE; Hull RL, Utzschneider KM (2006) Mechanism linking obesity to insulin resistance and type 2 diabetes. Nature 444: 840-846.

Han J., Lee Y., Yeom KH., Kim YH., Jim H., and Kim VN. The Drosha-DGCR8 in primary miRNA processing. Genes Dev. 2004; 18: 3016-3027

Hannun YA, Obeid LM (July 2002). "The Ceramide-centric universe of lipid-mediated cell regulation: stress encounters of the lipid kind". J. Biol. Chem. 277 (29): 25847–50

Hannun YA, Luberto C, Agraves KM (2001) Enzymes of sphingolipid metabolism: from modular to integrative signalling Biochemistry 40: 4893-4903

Hannun, Y. A. (1994) J. Biol. Chem. 269, 3125–3128

Harzer K, Paton BC, Poulos A, Kustermann-Kuhn B, Roggendorf W, Grisar T, Popp M (1989) Sphingolipid activator protein deficiency in a 16-week-old atypical Gaucher disease patient and his fetal sibling: biochemical signs of combined sphingolipidoses. Eur J Pediatr 149: 31-9

Hay, N. Sonenberg, N (2004) Upstream and downstream of mTOR. Gene Dev 18: 1926-1945.

Henschel A, Buchholz F, Habermann B (2004). "DEQOR: a web-based tool for the design and quality control of siRNAs". *Nucleic Acids Res* 32 (Web Server issue): W113–20.

Henseler M, Klein A, Reber M, Vanier MT, Landrieu P, Sandhoff K (1996) Analysis of a splice-site mutation in the sap-precursor gene of a patient with metachromatic leukodystrophy. Am J Hum Genet 58: 65-74

Hiesberger T, Hüttler S, Rohlmann A, Schneider W, Sandhoff K, Herz J. 1998. Cellular uptake of saposin (SAP) precursor and lysosomal delivery by the low density lipoprotein receptor-related protein (LRP). *EMBO J.* 17:4617–25

Hiraiwa M, Campana WM, Mizisin AP, Mohiuddin L, O'Brien JS (1999) Prosaposin: a myelinotrophic protein that promotes expression of myelin constituents and is secreted after nerve injury. Glia 26: 353-60

Ho MW, O'Brien JS. 1971. Gaucher's disease: deficiency of 'acid' β-glucosidase and reconstitution of enzyme activity in vitro. *Proc. Natl. Acad. Sci. USA* 68:2810–2813.

Holtschmidt H, Sandhoff K, Kwon HY, Harzer K, Nakano T, Suzuki K (1991) Sulfatide activator protein. Alternative splicing that generates three mRNAs and a newly found mutation responsible for a clinical disease. J Biol Chem 266: 7556-60

Hu L, Wang Z, Hu C, Liu X, Yao L, Li W, Qi Y (2005). "Inhibition of Measles virus multiplication in cell culture by RNA interference". *Acta Virol* 49 (4): 227–34.

Huesken D, Lange J, Mickanin C, Weiler J, Asselbergs F, Warner J, Meloon B, Engel S, Rosenberg A, Cohen D, Labow M, Reinhardt M, Natt F, Hall J (2005). "Design of a genome-wide siRNA library using an artificial neural network". *Nat Biotechnol* 23 (8): 995–1001

Hulkova H, Cervenkova M, Ledvinova J, Tochackova M, Hrebicek M, Poupetova H, Befekadu A, Berna L, Paton BC, Harzer K, Boor A, Smid F, Elleder M (2001) A novel mutation in the coding region of the prosaposin gene leads to a complete deficiency of prosaposin and saposins, and is associated with a complex sphingolipidosis dominated by lactosylceramide accumulation. Hum Mol Genet 10: 927-940

Ichikawa, S. et al. (1994) Proc. Natl. Acad. Sci. U. S. A. 91, 2703–2707

Ichikawa, S. et al. (1996) Proc. Natl. Acad. Sci. U. S. A. 93, 4638–4643

Ito, M. and Komori, H. (1997) Exp. Med. 15, 1476–1482

Izquierdo M (2005). "Short interfering RNAs as a tool for cancer gene therapy". *Cancer Gene Ther* 12 (3): 217–27.

Jakymiw A, Lian S, Eystathioy T, Li S, Satoh M, Hamel J, Fritzler M, Chan E (2005). "Disruption of P bodies impairs mammalian RNA interference". *Nat Cell Biol* 7 (12): 1267–1274.

Janitz M, Vanhecke D, Lehrach H (2006). "High-throughput RNA interference in functional genomics". *Handb Exp Pharmacol* 173: 97–104.

Jatzkewitz H, Stinshoff K (1973) An activator of cerebroside sulphatase in human normal liver and in cases of congenital metachromatic leukodystrophy. FEBS Lett 32: 129-31

Jeckel, D. et al. (1992) J. Cell Biol. 117, 259–267

Jia F, Zhang Y, Liu C (2006). "A retrovirus-based system to stably silence hepatitis B virus genes by RNA interference". *Biotechnol Lett* 28 (20): 1679–85.

Jiang M, Milner J (2002). "Selective silencing of viral gene expression in HPV-positive human cervical carcinoma cells treated with siRNA, a primer of RNA interference". *Oncogene* 21 (39): 6041–8

Kamath R, Ahringer J (2003). "Genome-wide RNAi screening in Caenorhabditis elegans". *Methods* 30 (4): 313–21.

Kennerdell J. R. and Carthew. RW. (2000), heritable gene silencing in *Drosophila* using dsRNA, *Nat. Biotechnol, Vol.* 18, pp. 896-898.

Kishimoto Y, Hiraiwa M, O'Brien JS (1992) Saposins: structure, function, distribution, and molecular genetics. J Lipid Res 33: 1255-67

Kolesnick, R. and Golde, D. W. (1994) Cell 77, 325–328

Kolter T, Sandhoff K (2005) Principles of lysosomal membrane digestion: stimulation of sphingolipid degradation by sphingolipid activator proteins and anionic lysosomal lipids. Annu Rev Cell Dev Biol 21: 81-103

Kolter T, Winau F, Schaible UE, Leippe M, Sandhoff K (2005) Lipid-binding proteins in membrane digestion, antigen presentation, and antimicrobial defense. J Biol Chem 280: 41125-8

Kretz KA, Carson GS, Morimoto S, Kishimoto Y, Fluharty AL, O'Brien JS. 1990. Characterization of a mutation in a family with saposin B deficiency: a glycosylation site defect. *Proc. Natl. Acad. Sci. USA* 87:2541–44

Kusov Y, Kanda T, Palmenberg A, Sgro J, Gauss-Müller V (2006). "Silencing of hepatitis A virus infection by small interfering RNAs". *J Virol* **80** (11): 5599–610.

Lahiri S and Futerman AH (2007) The metabolism and function of sphingolipids and glycosphingolipids. Cell. Mol Life Sci 64: 2270-2284

Lahiri S.and Futerman A.H., Lass5 is a bona fide dihydroceramide synthase that selectively utilizes palmitoyl-CoA as acyl donor, J. Biol. Chem. **280** (2005), pp. 33735–33738.

Lannert, H. et al. (1994) FEBS Lett. 342, 91–96

Laviad E.L., Albee L., Pankova-Kholmyansky I., Epstein S., Park H., Merrill Jr A.H., and Futerman A.H., Characterization of ceramide synthase 2: tissue distribution, substrate specificity, and inhibition by sphingosine 1-phosphate, J. Biol. Chem. **283** (2008), pp. 5677–5684.

Lavie, Y. et al. (1996) J. Biol. Chem. 271, 19530–19536

Lavie, Y. et al. (1997) J. Biol. Chem. 272, 1682–1687

Lee Y, Ahn C, Han J, Choi H, Kim J, Yim J, Lee J, Provost P. The nuclear RNAse III Drosha initiates miRNA processing. Nature 2003; 425: 415-419

Lee Y, Nakahara K, Pham J, Kim K, He Z, Sontheimer E, Carthew R (2004). "Distinct roles for Drosophila Dicer-1 and Dicer-2 in the siRNA/miRNA silencing pathways". *Cell* **117** (1): 69–81.

Lee Y., Hur I., Park SY., Kim YK., Suh MR., Kim VN. The role of PACT in the RNAi pathway. EMBO J 2006; 25: 522-532

Leuschner P, Ameres S, Kueng S, Martinez J (2006). "Cleavage of the siRNA passenger strand during RISC assembly in human cells". *EMBO Rep* **7** (3): 314–3120.

Li LC (2008). "Small RNA-Mediated Gene Activation". *RNA and the Regulation of Gene Expression: A Hidden Layer of Complexity*. Caister Academic Press. ISBN 978-1-904455-25-7.

Li SC, Sonnino S, Tettamanti G, Li YT. 1988. Characterization of a nonspecific activator protein for the enzymatic hydrolysis of glycolipids. *J. Biol. Chem.* 263:6588–91

Li SC, Kihara H, Serizawa S, Li YT, Fluharty AL, Mayes JS, Shapiro LJ. 1985. Activator protein required for the enzymatic hydrolysis of cerebroside sulfate. Deficiency in urine of patients affected with cerebroside sulfatase activator deficiency and identity with activators for the enzymatic hydrolysis of GM1 ganglioside and globotriaosylceramide. *J. Biol. Chem.* 260:1867–1871.

Li C, Parker A, Menocal E, Xiang S, Borodyansky L, Fruehauf J (2006). "Delivery of RNA interference". *Cell Cycle* **5** (18): 2103–9.

Linke T, Wilkening G, Sadeghlar F, Mozcall H, Bernardo K, et al. 2001b. Interfacial regulation of acid ceramidase activity. Stimulation of ceramide degradation by lysosomal lipids and sphingolipid activator proteins. *J. Biol. Chem.* 276:5760–5768

Liu Q, Rand T, Kalidas S, Du F, Kim H, Smith D, Wang X (2003). "R2D2, a bridge between the initiation and effector steps of the Drosophila RNAi pathway". *Science* **301** (5641): 1921–1925.

Lodish H, Berk A, Matsudaira P, Kaiser CA, Krieger M, Scott MP, Zipurksy SL, Darnell J (2004). *Molecular Cell Biology*. WH Freeman: New York, NY.

Lefrancois S, Zeng J, Hassan AJ, Canuel M, Morales CR. 2003. The lysosomal trafficking of sphingolipid activator proteins (SAPs) is mediated by sortilin. *EMBO J.* 22:6430–37

Linke T, Wilkening G, Lansmann S, Moczall H, Bartelsen O, et al. 2001a. Stimulation of acid sphingomyelinase activity by lysosomal lipids and sphingolipid activator proteins. *Biol. Chem.* 382:283–90

Linke T, Wilkening G, Sadeghlar F, Mozcall H, Bernardo K, et al. 2001b. Interfacial regulation of acid ceramidase activity. Stimulation of ceramide degradation by lysosomal lipids and sphingolipid activator proteins. *J. Biol. Chem.* 276:5760–5768

Liu J, Carmell MA, Rivas FV, Marsden CG, Thomson JM, Song JJ, Hammond SM, Joshua-Tor L. Argonaute2 is the catalytic engine of mammalian RNAi. Science 2004; 305: 1437-41.

Macrae I, Zhou K, Li F, Repic A, Brooks A, Cande W, Adams P, Doudna J (2006). Structural basis for double-stranded RNA processing by dicer. *Science* **311** (5758): 195–198.

Matranga C, Tomari Y, Shin C, Bartel D, Zamore P (2005). "Passenger-strand cleavage facilitates assembly of siRNA into Ago2-containing RNAi enzyme complexes". *Cell* **123** (4): 607–620.

Matsuda J, Kido M, Tadano-Aritomi K, Ishizuka I, Tominaga K, et al. 2004. Mutation in saposin D domain of sphingolipid activator protein gene causes urinary system defects and cerebellar Purkinje cell degeneration with accumulation of hydroxy fatty acid-containing ceramide in mouse. *Hum. Mol. Genet.* 13:2709–23.

Matsuda J, Vanier MT, Saito Y, Tohyama J, Suzuki K (2001) A mutation in the saposin A domain of the sphingolipid activator protein (prosaposin) gene results in a late-onset, chronic form of globoid cell leukodystrophy in the mouse. Hum Mol Genet 10: 1191-9

McGinnis K, Chandler V, Cone K, Kaeppler H, Kaeppler S, Kerschen A, Pikaard C, Richards E, Sidorenko L, Smith T, Springer N, Wulan T (2005). "Transgene-induced RNA interference as a tool for plant functional genomics". *Methods Enzymol* **392**: 1–24.

Mehl E, Jatzkewitz H. 1964. A cerebrosidesulfatase from swine kidney. *Hoppe Seylers Z. Physiol. Chem.* 339:260–76

Meister G, Landthaler M, Patkaniowska A, Dorsett Y, Teng G, Tuschl T. Human Argonaute 2 mediates RNA cleavage targeted by miRNAs and siRNAs. Mol Cell 2004; 15: 185-197.

Merrill. "Characterization of serine palmitoyltransferase activity in Chinese hamster ovary cells." Biochim Biophys Acta (1983) 754(3):284-91.

Merrill and Williams. "Utilization of different fatty acyl-CoA thioesters by serine palmitoyltransferase from rat brain". Journal of Lipid Research (1984) 25 (2): 185-188.

Myers, J. W., Jones, J. T., Meyer, T. et al. (2003), 'Recombinant Dicer efficiently converts large dsRNAs into siRNAs suitable for gene silencing', Nat. Biotechnol., Vol. 21, pp. 324–328.

Mizutani, A. Kihara and Y. Igarashi, Lass3 (longevity assurance homologue 3) is a mainly testis-specific (dihydro) ceramide synthase with relatively broad substrate specificity, Biochem. J. **398** (2006), pp. 531–538.

Mizutani Y., Kihara A. and Igarashi Y., Mammalian lass6 and its related family members regulate synthesis of specific ceramides, Biochem. J. 390 (2005), pp. 263–271.

Morita T, Mochizuki Y, Aiba H (2006). "Translational repression is sufficient for gene silencing by bacterial small noncoding RNAs in the absence of mRNA destruction". *Proc Natl Acad Sci USA* **103** (13): 4858–63.

Mraz M, Pospisilova S, Malinova K, *et al.* (March 2009). "MicroRNAs in chronic lymphocytic leukemia pathogenesis and disease subtypes". *Leuk Lymphoma* **50** (3): 506–9.

Naito Y, Ui-Tei K, Nishikawa T, Takebe Y, Saigo K (2006). "siVirus: web-based antiviral siRNA design software for highly divergent viral sequences". *Nucleic Acids Res* **34** (Web Server issue): W448–50.

Naito Y, Yamada T, Ui-Tei K, Morishita S, Saigo K (2004). "siDirect: highly effective, target-specific siRNA design software for mammalian RNA interference". *Nucleic Acids Res* **32** (Web Server issue): W124–9.

Naito Y, Yamada T, Matsumiya T, Ui-Tei K, Saigo K, Morishita S (2005). Check: highly sensitive off-target search software for "double-stranded RNA-mediated RNA interference". *Nucleic Acids Res* **33** (Web Server issue): W589–91.

Nakayashiki H, Kadotani N, Mayama S (2006). "Evolution and diversification of RNA silencing proteins in fungi". *J Mol Evol* **63** (1): 127–35.

Nozue, M. et al. (1988) Int. J. Cancer 42, 734–738

Okamura K, Ishizuka A, Siomi H, Siomi M (2004). "Distinct roles for Argonaute proteins in small RNA-directed RNA cleavage pathways". *Genes Dev* **18** (14): 1655–1666.

O'Brien JS, Kishimoto Y (1991) Saposin proteins: structure, function, and role in human lysosomal storage disorders. Faseb J 5: 301-308.

Paddison P, Caudy A, Hannon G (2002). "Stable suppression of gene expression by RNAi in mammalian cells". *Proc Natl Acad Sci USA* **99** (3): 1443–8.

Paddison, P. J., Caudy, A. A., Bernstein, E. (2002), 'Short hairpin RNAs (shRNAs) induce sequence-specific silencing in mammalian cells', Genes Dev., Vol. 16, pp. 948–958.

Paddison, P. J. and Hannon, G. J. (2002), 'RNA interference: the new somatic cell genetics?' Cancer Cell, Vol. 2, pp. 17–23.

Pak J, Fire A (2007). "Distinct populations of primary and secondary effectors during RNAi in C. elegans". *Science* **315** (5809): 241–244.

Patel RT, Soulages JL, Hariharasundaram B, Arrese L (2005) Activation of the lipid droplet controls the rate of lipolysis of triacylglycerides in the insect fat body. *J. Biol Chem.* **280**: 22624-22631.

Pewzner-Jung et al. "When do Lasses (longevity assurrance genes) become CerS (ceramide - Synthases)? Insights into the regulation of ceramide synthesis". Journal of Biological Chemistry. (2006) 281, 2500125005.

Pillai RS, Bhattacharyya SN, Filipowicz W. "Repression of protein synthesis by miRNAs: how many mechanisms?". *Trends Cell Biol.*

Platt FM, Walkley SU. 2004. Lysosomal defects and storage. In *Lysosomal Disorders of the Brain*, ed. FM Platt, SU Walkley, pp. 32–49. New York: Oxford Univ. Press Salvioli R, Tatti M, Ciaffoni F, Vaccaro AM. 2000. Further studies on the reconstitution of glucosylceramidase activity by Sap C and anionic phospholipids. *FEBS Lett.* 472:17–21

Preall J, He Z, Gorra J, Sontheimer E (2006). "Short interfering RNA strand selection is independent of dsRNA processing polarity during RNAi in Drosophila". *Curr Biol* **16** (5): 530–535.

Pruett et al. "Biodiversity of sphingoid bases ("sphingosines") and related amino alcohols". Journal of Lipid Research. (2008) 49:1621-1639.

Putral L, Gu W, McMillan N (2006). "RNA interference for the treatment of cancer". *Drug News Perspect* **19** (6): 317–24

Qi X, Grabowski GA (2001) Differential membrane interactions of saposins A and C: implications for the functional specificity. J Biol Chem 276: 27010-7

Qi X, Qin W, Sun Y, Kondoh K, Grabowski GA (1996) Functional organization of saposin C. Definition of the neurotrophic and acid beta-glucosidase activation regions. J Biol Chem 271: 6874-6880

Qiu S, Adema C, Lane T (2005). "A computational study of off-target effects of RNA interference". *Nucleic Acids Res* **33** (6): 1834–1847.

Rafi MA, Zhang XL, DeGala G, Wenger DA (1990) Detection of a point mutation in sphingolipid activator protein-1 mRNA in patients with a variant form of metachromatic leukodystrophy. Biochem Biophys Res Commun 166: 1017-23

Raoul C, Barker S, Aebischer P (2006). "Viral-based modelling and correction of neurodegenerative diseases by RNA interference". *Gene Ther* **13** (6): 487–95.

Rende M, Brizi E, Donato R, Provenzano C, Bruno R, Mizisin AP, Garrett RS, Calcutt NA, Campana WM, O'Brien JS (2001) Prosaposin is immunolocalized to muscle and prosaptides promote myoblast fusion and attenuate loss of muscle mass after nerve injury. Muscle Nerve 24: 799-808

Reynolds A, Anderson E, Vermeulen A, Fedorov Y, Robinson K, Leake D, Karpilow J, Marshall W, Khvorova A (2006). "Induction of the interferon response by siRNA is cell type- and duplex length-dependent". *RNA* **12** (6): 988–93.

Riebeling C., Allegood J.C., Wang E., Merrill A.H. Jr. and Futerman A.H.. Two mammalian longevity assurance gene (lag1) family members, trh1 and trh4, regulate dihydroceramide synthesis using different fatty acyl-CoA donors, J. Biol. Chem. **278** (2003), pp. 43452–43459.

Robinson K, Beverley S (2003). "Improvements in transfection efficiency and tests of RNA interference (RNAi) approaches in the protozoan parasite Leishmania". *Mol Biochem Parasitol* **128** (2): 217–28.

Rosen ED, Spiegelmann BM (2006) Adipocytes as regulators of energy balance and glucose homeostasis. Nature 444: 847-853.

Simba V, Garg A (2006) Lipodystrophy: lessons in lipid and energy metabolis. Curr Opin Lipidol 17: 162-169

Sah D (2006). "Therapeutic potential of RNA interference for neurological disorders". *Life Sci* **79** (19): 1773–80.

Sando, G. N., Howard, E. J. and Madison, K. C. (1996) J. Biol.Chem. 271, 22044–22051

Sandhoff K, Kolter T, Harzer K. 2001. Sphingolipid activator proteins. In *The Metabolic and Molecular Bases of Inherited Disease*, ed. CR Scriver, AL Beaudet, WS Sly, D Valle, III:3371–88. New York: McGraw-Hill. 8th ed.

Schepers, U. and T. Kolter (2001). "RNA interference: A new way to analyze protein function Angewandte Chemie-International Edition 40(13): 2437

Schnabel D, Schröder M, Sandhoff K. 1991. Mutation in the sphingolipid activator protein 2 in a patient with a variant of Gaucher disease. *FEBS Lett.* 284:57–59

Schulte, S. and Stoffel, W. (1993) Proc. Natl. Acad. Sci.

Schwarz DS, Hutwagner G, Du T, Zu Z, Aronin N, Zamore PD. Assymetry in the assembly of the RNAi silencing complex. Cell 2003; 115: 199-2008.

Sijen T, Steiner F, Thijssen K, Plasterk R (2007). "Secondary siRNAs result from unprimed RNA synthesis and form a distinct class". *Science* **315** (5809): 244–247.

Siomi, Haruhiko; Siomi, Mikiko C. (22 January 2009). "On the road to reading the RNA-interference code". *Nature* 457 (7228): 396–404.

Soeda S, Hiraiwa M, O'Brien JS, Kishimoto Y (1993) Binding of cerebrosides and sulfatides to saposins A-D. J Biol Chem 268: 18519-23

Spassieva S., Seo J.G.,. Jiang J.C, Bielawski J., Alvarez-Vasquez F., Jazwinski S.M., Hannun Y.A. and Obeid L.M., Necessary role for the lag1p motif in (dihydro)ceramide synthase activity, *J. Biol. Chem.* **281** (2006), pp. 33931–33938.

Spiegel R, Bach G, Sury V, Mengistu G, Meidan B, et al. 2005. A mutation in the saposin A coding region of the prosaposin gene in an infant presenting as Krabbe disease: first report of saposin A deficiency in humans. *Mol. Genet. Metab.* 84:160–66

Spiegel S, Milstien S (July 2002). Sphingosine 1-phosphate, a key cell signaling molecule. *J. Biol. Chem.* **277** (29): 25851–4.

Spiegel S. and Milstien S. (2000) Sphingosin-1-phosphate: singnaling inside and out. FEBS Lett 30: 55-57

Stahl, N. et al. (1994) J. Neurosci. Res. 38, 234–242

Stein P, Zeng F, Pan H, Schultz R (2005). "Absence of non-specific effects of RNA interference triggered by long double-stranded RNA in mouse oocytes". *Dev Biol* **286** (2): 464–71.

Sun Y, Qi X, Grabowski GA. 2003. Saposin C is required for normal resistance of acid β-glucosidase to proteolytic degradation. *J. Biol. Chem.* 278:31918–23

Sun Y, Qi X, Witte DP, Ponce E, Kondoh K, Quinn B, Grabowski GA (2002) Prosaposin: threshold rescue and analysis of the "neuritogenic" region in transgenic mice. Mol Genet Metab 76: 271-86

Suzuki K, Vanier MT. 1999. Lysosomal and peroxisomal diseases. In *Basic Neurochemistry—Molecular, Cellular and Medical Aspects*, ed. GJ Siegel, BW Agranoff, RW Albers, SK Fisher, MD Uhler, pp. 821–839.

Takeshita F, Ochiya T (2006). "Therapeutic potential of RNA interference against cancer". *Cancer Sci* **97** (8): 689–96.

Tang G. miRNA and siRNA: an insight into RISCs. Trends Biochem Sci. 2005; 30: 106-114.

Teufel A., Maass T., Galle P.R.and Malik N. The longevity assurance homologue of yeast lag1 (lass) gene family (review), *Int. J. Mol. Med.* **23** (2009), pp. 135–140.

Timmons L, Court DL and Fire A (2001). Injestion of bacterially expressed dsRNA can induce potent and specific genetic interference in *C. elegangs*. Gene, Vol 263; pp. 103-112.

Timmons L., Tabraa H., and Mello CC. (2003) Inducible systemic RNA silencing in *C. elegans*. *Mol. biol. Cell* Vol. 14, pp, 2972- 2983

Tiscornia G, Tergaonkar V, Galimi F, Verma I (2004). "CRE recombinase-inducible RNA interference mediated by lentiviral vectors". *Proc Natl Acad Sci USA* **101** (19): 7347–51.

Tomari Y, Matranga C, Haley B, Martinez N, Zamore P (2004). "A protein sensor for siRNA asymmetry". *Science* **306** (5700): 1377–1380.

Tong A, Zhang Y, Nemunaitis J (2005). "Small interfering RNA for experimental cancer therapy". *Curr Opin Mol Ther* **7** (2): 114–24.

Travella S, Klimm T, Keller B (2006). "RNA interference-based gene silencing as an efficient tool for functional genomics in hexaploid bread wheat". *Plant Physiol* **142** (1): 6–20.

Vaccaro AM, Ciaffoni F, Tatti M, Salvioli R, Barca A, et al. 1995. pH-dependent conformational properties of saposins and their interactions with phospholipid membranes. *J. Biol. Chem.* 270:30576–80

Vaccaro AM, Salvioli R, Tatti M, Ciaffoni F (1999) Saposins and their interaction with lipids. Neurochem Res 24: 307-14

van Meer G, Lisman Q (July 2002). Sphingolipid transport: rafts and translocators. *J. Biol. Chem.* **277** (29): 25855–8.

van Weely S, Brandsma M, Strijland A, Tager JM, Aerts JM. 1993. Demonstration of the existence of a second, non-lysosomal glucocerebrosidase that is not deficient in Gaucher disease. *Biochim. Biophys. Acta* 1181:55–62

Vanhecke D, Janitz M (2005). "Functional genomics using high-throughput RNA interference". *Drug Discov Today* **10** (3): 205–12.

Ventura A, Meissner A, Dillon C, McManus M, Sharp P, Van Parijs L, Jaenisch R, Jacks T (2004). "Cre-lox-regulated conditional RNA interference from transgenes". *Proc Natl Acad Sci USA* **101** (28): 10380–5.

Vermeulen A, Behlen L, Reynolds A, Wolfson A, Marshall W, Karpilow J, Khvorova A (2005). "The contributions of dsRNA structure to dicer specificity and efficiency". *RNA* **11** (5): 674–682.

Vogel A, Schwarzmann G, Sandhoff K. 1991. Glycosphingolipid specificity of the human sulfatide activator protein. *Eur. J. Biochem.* 200:591–597

Volpe T, Schramke V, Hamilton G, White S, Teng G, Martienssen R, Allshire R (2003). "RNA interference is required for normal centromere function in fission yeast". *Chromosome Res* **11** (2): 137–146.

Waterhouse P. M. and Helliwell C. A. (2003), Exploring plant genomes by RNA induced gene silencing, *Nat. Rev. Gen.,* Vol 4., pp. 29-38

Williams, B. R. (1997), 'Role of the double-stranded RNA-activated protein kinase (PKR) in cell regulation', Biochem. Soc.Trans., Vol. 25, pp. 509–513.

Wilkening G, Linke T, Sandhoff K. 1998. Lysosomal degradation on vesicular membrane surfaces. Enhanced glucosylceramide degradation by lysosomal anionic lipids and activators. *J. Biol. Chem.* 273:30271–78

Winau F, Schwierzeck V, Hurwitz R, Remmel N, Sieling PA, Modlin RL, Porcelli SA, Brinkmann V, Sugita M, Sandhoff K, Kaufmann SH, Schaible UE (2004) Saposin C is required for lipid presentation by human CD1b. Nat Immunol 5: 169-74

Winchester BG. 2004. Primary defects in lysosomal enzymes. In *Lysosomal Disorders of the Brain*, ed. FM Platt, SU Walkley, pp. 81–130. New York: Oxford Univ. Press.

Yu, J. Y., DeRuiter, S. L. and Turner, D. L. (2002), 'RNA interference by expression of short-interfering RNAs and hairpin RNAs in mammalian cells', Proc. Natl. Acad. Sci. USA,Vol. 99, pp. 6047–6052.

Yuan Y J, Meister G, Pei Y, Tuschl T, Patel D (2005). "Structural basis for 5'-end-specific recognition of guide RNA by the A. fulgidus Piwi protein". *Nature* **434** (7033): 666–6670

Zhao Y, Srivastava D (2007). "A developmental view of microRNA function". *Trends Biochem. Sci.* **32** (4): 189–197.

Zamore PD, Tuschl T, Sharp PA, Bartel D (2000). "RNAi: double-stranded RNA directs the ATP-dependent cleavage of mRNA at 21 to 23 nucleotide intervals". *Cell* **101** (1): 25–33.

Zang H., Kolb FA., Jaskeevicz L., Westhof E., Filipovicz W. Single processing center models fonr human Dicer and bacterial RNA III. Cell 2004; 118: 57-68

Zechner R, Strauss JK, Haemmerle G, Lass A, Zimmermann R (2005) Lipolysis: pathway under construction. Curr Opin Lipidol 16: 333-340.

Zender L, Hutker S, Liedtke C, Tillmann H, Zender S, Mundt B, Waltemathe M, Gosling T, Flemming P, Malek N, Trautwein C, Manns M, Kuhnel F, Kubicka S (2003). "Caspase 8 small interfering RNA prevents acute liver failure in mice". *Proc Natl Acad Sci USA* **100** (13): 7797–802.

Zitomer NC, Mitchell T, Voss KA, Bondy GS, Pruett ST, Garnier-Amblard EC, Liebeskind LS, Park H, Wang E, Sullards MC, Merrill AH Jr, Riley RT. "Ceramide Synthase Inhibition by Fumonisin B1 Causes Accumulation of 1-Deoxysphinganine: A Novel Category of Bioactive 1-Deoxysphingoid Bases And 1-Deoxydihydroceramides Biosynthesized By Mammalian Cell Lines And Animals". Journal of Biological Chemistry (2009) 284 (8): 4786-4795.